The third-order Fresnel lens from Cape Spencer Lighthouse, which has been at the Alaska State Museum since its removal in 1974. (Sara Boesser)

A SHORT, BRIGHT FLASH

AUGUSTIN FRESNEL AND THE BIRTH OF THE MODERN LIGHTHOUSE

THERESA LEVITT

W. W. NORTON & COMPANY

NEW YORK · LONDON

For information about permission to reproduce selections from this book, write to Permissions, W. W. Norton & Company, Inc., 500 Fifth Avenue, New York, NY 10110

For information about special discounts for bulk purchases, please contact W. W. Norton Special Sales at specialsales@wwnorton.com or 800-233-4830

Manufacturing by Friesens
Production manager Leeann Graham

Cartography by George Ward
Book design by Robert L. Wiser
Composed in Peccadillo, Engraver's Gothic, and Perpetua

Library of Congress Cataloging-in-Publication Data

Levitt, Theresa.
    A short bright flash: Augustin Fresnel and the birth of the modern lighthouse / Theresa Levitt.—First edition.
        pages   cm
    Includes bibliographical references and index.
    ISBN 978-0-393-06879-5 (hbk.)
    1. Fresnel, Augustin Jean, 1788–1827. 2. Lighthouses—History. 3. Fresnel lenses—History. 4. Physicists—France—Biography. I. Title. II. Title: Augustin Fresnel and the birth of the modern lighthouse.
    VK1015.L48 2013
    623.89'42—dc23
    [B]
                                                            2013001565

W. W. Norton & Company, Inc., 500 Fifth Avenue, New York, N.Y. 10110
    www.wwnorton.com
W. W. Norton & Company Ltd., Castle House, 75/76 Wells Street, London W1T 3QT

0 9 8 7 6 5 4 3 2 1

Illustration Credits: 17 Alinari/Art Resource; 57 Thomas Tag; 119 Erich Lessing/Art Resource, NY; 191 Courtesy Outer Banks History Center, Manteo, North Carolina; 223 Thomas Tag.

CONTENTS

PREFACE

THE FIRST FRESNEL LENS I ever saw was at the Alaska State Museum. I went there frequently as a child and later worked the front desk as a high school student. It was a great museum. The first floor was given over to natural history and native cultures, while the top floor was the historical section. You reached the historical section by walking up a long ramp that circled around an enormous spruce tree containing a nest of taxidermic bald eagles, complete with a continuous audio loop playing their surprisingly high-pitched cries. As you turned the final corner of the ramp, the first thing you saw was the lens. It was almost impossible not to look at the strange, glittering contraption. The hundreds of finely polished prisms placed into a precise yet inscrutable pattern gave it an unworldly effect, especially when you have just emerged from the wilds of the boreal forest. The unusual design of the lens, including its sharply peaked top, made it look like a spaceship that had accidentally touched down in Alaska. Interference patterns played continually across its prisms, adding the shimmering palette of a soap bubble to the already gleaming structure. With its sleek geometries, thick glass crystals, and brass frame, the lens was equal parts Victorian fussiness and science fiction. The nineteenth century had met the future, producing this steampunk device that seemed completely, utterly incongruous amid the skunk cabbage, old man's beard, and eagle calls.

Because it had been placed at the entrance to the historical section, I had always associated the lens with the dawning of some new era in Alaskan history. Even at the time I knew this was patently naive, and that its location in the museum surely had more to do with floor plans and traffic flow than historical accuracy. As I began to study the origins of the Fresnel lens, however, I came to realize that it did mark a new era of sorts. While it was true that people had been fishing, hunting, and trading on Alaska's coasts for centuries, these were essentially local endeavors, and few people outside of the district paid much attention. In fact, Americans had derided the purchase of the Alaskan territory as "Seward's Folly" precisely because it seemed to have little that could be commoditized for growing world trade. The discovery of gold in the Klondike changed that situation, and the decision to buy eleven Fresnel lenses for the Alaskan coast followed within weeks. Although the complexity of construction delayed the lighting of the shores until after the gold rush was over, the lenses did contribute to the shipping lanes becoming more secure and vastly more traveled, and marked Alaska's entrance into the realm of global commerce. Everywhere I looked, the story repeated itself. The moment a Fresnel lens appeared at a location was the moment that region became linked into the world economy. And just like that, the lens became a "lens," my own particular optic for bringing the past into focus. The metaphor has been used frequently in history, and to be honest, I have often been annoyed when someone said they're going to use some sort of thing or another as "a lens" to view history. It's not that I don't like the optical analogy (as a historian of optics, there are few things I like more), but it seemed imprecise. Lenses, after all, can do a lot of different things. They can magnify, telescope, invert, diverge, converge, and correct, all while inevitably distorting the light that you see. I never knew exactly which one of these the metaphor meant, yet I was soon the most enthusiastic of converts. The more intently I stared at the shiny glass prisms, the more clearly the contours of the past two centuries took form.

I followed my lenses from the French Revolution to the Second World War, a period given over to baggy amorphous categories like modernity, nationalism, and empire. The era marked the modern age of globalization, defined by the rapid development of a truly global transportation infrastructure and sharp intensification of economic interdependence. Any attempts to narrow my focus for the sake of manageability were destined to leave out the complex flows and international networks that made this period exciting. Studying the Fresnel lens did not narrow anything at all, but somehow made clearer the far-flung reach of this period's expansion, with each light serving as a literal beacon of the new global world.

With its elegant engineering and sleek lines, the Fresnel lens is as much an icon of modernity as the Eiffel Tower. It too was a star of international exhibitions, where crowds in the millions lined up to marvel at its newness and power. But more than an icon, the Fresnel lens participated directly in the creation of this era. Safely lighting the seaways of human commerce allowed international trade and colonial ambitions to flourish. Harnessing the laws of optics to save lives provided a capital argument for the nineteenth century's vision of science as the engine of progress and civilization. At the dawn of this age, science enthusiast William Blake exhorted us "to see a world in a grain of sand," and so let's mix that sand with a bit of ash, heat it in a crucible, and shape it into our lens for seeing the world.

DARK AND DEADLY SHORES

ONE STORY ALONE dominated newspaper headlines in the autumn of 1817: the tragic, grisly fate of the passengers and crew on the French frigate *la Méduse*. As lurid new details emerged from the trial of the ship's court-martialed captain, vicomte Hugues Duroy de Chaumareys, and from the published account by two survivors, Parisians could talk of little else. The frigate had run aground on the Bank of Arguin, off the west coast of Africa, on July 2, 1816. With too little space in the lifeboats for everyone aboard, the crew cobbled together a makeshift, twenty-by-sixty-foot raft of planks and broken masts. Leaving a few crew members with the ship, de Chaumareys directed one woman and 146 men onto the raft, mostly soldiers, both French and mercenary, but also colonists bound for Senegal, and a few of the ship's officers. He promised, loudly and frequently, that the more seaworthy lifeboats would tow the raft to shore. The operation proved difficult, however, and the occupants of the raft soon saw the crew on the captain's lifeboat hacking at the towrope with a hatchet and crying "We abandon them!" The lifeboats rapidly disappeared from sight, leaving the desperately overcrowded raft at the mercy of the currents. With all but the barest of provisions

*"We abandon them!" The moment when the captain's boat
cut the towrope to the raft.*

jettisoned, the raft still sat dangerously below the water's
surface. Many in the tightly huddled group were submerged
up to their waists, and even those sitting on the highest por-
tions were soaked by waves passing over.

No one knows exactly how many people died the first
night—most likely dozens. Some were swept overboard
during a violent storm; others were crushed between the
raft's wooden lattice. The morning brought calmer waters,
but heat and thirst began to take a toll. Delirium spread.
Three men threw themselves into the sea. Murder followed,
as a frenzied, mutinous riot pitted the survivors against one
another. Several soldiers smashed open one of a few remaining wine casks and, having drunk all they could, turned on the
small group of officers who constituted the raft's leaders. The
officers, outnumbered but better armed, returned in kind.
Sixty people died in the fighting, and nearly all of the meager
provisions were destroyed.

By dawn on the third day, there was no food left—that is, except for an ample supply of fresh corpses. The cannibalism began furtively. The next day, the entire raft joined in, in a communal meal that combined the corpses' flesh with a few flying fish that had become entangled in the raft's beams. From that point on, everyone ate human flesh without ceremony. Another round of fighting between the mercenaries and the officers during the night left yet more bodies. It was now easy to count the survivors: thirty, including the lone woman, a canteen manager for the army. The ship's doctor, J. B. Henry Savigny, who had emerged as the de facto leader, estimated that the amount of wine left would last them four days. Two soldiers discovered stealing it were thrown overboard. The youngest passenger, a twelve-year-old sailor, perished. On the seventh day, Savigny and the remaining officers agreed that the practical plan was to sacrifice the weakest of the group, to conserve resources and improve the chances of survival for the rest. He designated twelve people he thought likely to die within forty-eight hours, including the woman and her husband, a soldier. The remaining fifteen oversaw their execution and then threw their own guns into the water to prevent any further violence.

These fifteen men were still alive on July 17, when one of them, an infantry captain, spotted the topmasts of the *Argus*, a French brig passing by. By September, the survivors were back in Paris. Over the course of the next year, the full extent of the tragedy unfolded during the lengthy trial of Captain de Chaumareys, whose own lifeboat, after abandoning the raft, had proceeded to a French colonial outpost, Saint-Louis, in Senegal. (The passengers of the other lifeboats, meanwhile, were forced ashore for lack of provisions and hiked a grueling two hundred miles south through the Sahara to reach Saint-Louis.)

Of the seventeen men who had stayed with the wreck, three survived, starved and skeletal by the time a ship rescued them.

The public was captivated by the horrific events that occurred on the makeshift raft, soon revealed in every detail in a book by Savigny and another survivor, the geographer Alexandre Corréard. Their account was the latest in a long line of best-selling shipwreck stories, including fictional ones inspired by Daniel Defoe's *Robinson Crusoe* (1719), such as *The Swiss Family Robinson*, a children's book published shortly before *la Méduse* set sail. Few events weighed as heavily on the nineteenth-century imagination as a ship on the rocks.

Shipwrecks, a constant fact of life in those days, almost inevitably meant death, since most sailors didn't know how to swim and rescue services hardly existed. It is impossible to know exactly how many ships went down because governments did not start keeping track until the middle of the century, but evidence suggests that the numbers were high. The insurer Lloyd's of London reported on its books that 362 ships were "wrecked" or "missing" for the year 1816 alone. In France, two naturalists counted nearly a hundred French ships that disappeared every year just in the English Channel in the period 1817–1820.

A few ships undoubtedly were lost on the open sea, overwhelmed by a storm. The majority, however, met their end by running into the shore. It may have been unnerving to leave the safety of the coastline behind and see nothing but water on the horizon ahead, yet as every sailor knew, it was the land that could kill you.

Lighthouses seem like the obvious solution to the problem, but they were few and far between. No one expected dangerous spots to be marked. Even though the Bank of Arguin was a well-known hazard, not much of anything

Illustration from Le Chapeau de M. Lambert, *a story by Max de Rével of a ship that wrecks on its way to Martinique.*

Engraving of a shipwreck off the Île de Sein, in Brittany, France.

stood on this desolate edge of the Sahara. A single light on the coast could have told the crew of *la Méduse* where they were, offering them an opportunity to avoid disaster. But among the vicious and protracted finger-pointing that followed the debacle, no one complained about the lack of a light. Even the populated coasts of Europe remained mostly dark. France, a great maritime power, had a mere twenty lighthouses by 1800. England boasted a few more. No other country had more than a handful, and most of these were minor harbor lights, decidedly unimpressive affairs.

Lighthouses are reputedly one of the most ancient technologies. The Seven Wonders of the Ancient World include two built by the Greeks in the third century B.C.: the celebrated Pharos of Alexandria and the Colossus of Rhodes, an enormous statue that stood astride the harbor of Rhodes and held aloft a light. The Romans were enthusiastic lighthouse builders as well, leaving examples of their craft dotted along the shores of their former empire. These ancient accomplishments are misleading though, because they did not precisely fill the role that we now expect of lighthouses. They invariably stood at the entrances to harbors, primarily to guide ships in and out, not to warn them away.

Lighthouse technology changed little over the next two millennia. The light at Alexandria used a combination of flame and mirrors; at the dawn of the nineteenth century the technique was largely the same. Seventeenth- and eighteenth-century lighthouses usually marked ports. A few famous towers—Eddystone in England and Cordouan in France, for example—sat alone on top of dangerous rocks. The very extent of their renown, however, points to the unexpectedness of their purpose, and the feebleness of their lights was a frequent source of complaint. The short range of most lights, rarely more than a few miles, meant that by the time a ship saw them, it was already courting disaster. *La Méduse*, for example, ran aground more than four miles from shore. A light would have served only to lure the ship right into the sands.

In the autumn of 1817, as *la Méduse*'s story gripped Paris, a shy, awkward engineer named Augustin Fresnel arrived in the city. Not quite thirty years old, he had spent much of the previous decade in remote corners of France overseeing the building of roads while harboring secret dreams of finding glory in the nation's capital. Paris, in the first decades of the nineteenth century, was a magnet for the young, ambitious, and talented, looking to get ahead in the rapidly changing world.

That same autumn, another young man arrived in Paris with little to recommend him but big dreams. Painter Théodore Géricault, returning from an Italian tour, read the local headlines and conceived of a massive monumental work on the shipwreck of *la Méduse*. Visiting morgues and dissection theaters to

Théodore Géricault's The Raft of the Medusa.

gain a proper sense of the mutilated body, he painted the battered raft at the moment it spotted the *Argus* in the distance. The piece stood at more than sixteen feet high and twenty-three feet long, and its stark depiction of nature shocked and enthralled its audience. "The painter has assembled the most revolting aspects of despair, rage, hunger, death and putrefaction," wrote one reviewer (and this was one of the positive reviews).

Géricault presented *The Raft of the Medusa* at the Paris Salon that opened on August 25, 1819. That same week, on the opposite side of town, Fresnel made his own presentation to the assembled members of the Commission des phares, France's Lighthouse Commission, hazarding an idea for a new lens that would make lighthouses brighter. The Commission's response was a far cry from the buzz surrounding Géricault's painting, as its members offered little encouragement and admitted they did not see the point of the lens. They did tell Fresnel that if he made one (on his own time and at his own expense), they would look at it. That was enough encouragement for the young engineer. He set to work on his project—a masterpiece of glass and metal, rather than oil paint and canvas, that would itself come to capture the public's imagination.

Géricault's painting and Fresnel's lens were in many ways flip sides of one another, each presenting a different face of the nineteenth century's complex relationship with nature. Géricault's Romantic tableau cast nature as uncontrollable, and humans as being broken by its power. It invoked not beauty, but the Kantian "sublime"—a mixture of fear and delight that comes from nature surpassing our own ability to grasp it. Fresnel was equally fascinated by nature, but at the heart of his enterprise was the dream of control, literally bending light to his own desires. Nature's indomitable power, while not exactly tamed, would be funneled through the channels of science. If Géricault's painting provided its audiences a cathartic experience of the horror of a shipwreck, Fresnel's work offered the tantalizing hope that human progress might eradicate such scenes altogether.

Fresnel's initial idea was just the beginning of a story that winds through the major themes of the nineteenth century: an explosion in global trade, industrial technology, and overseas empire. The Fresnel lens, born at the intersection of empire, science, and engineering, fueled a revolution that transformed the nature of sea travel. Within a hundred years, more than ten thousand Fresnel lenses lined shores around the globe. Any unmarked hazard was now viewed as an assault on humanity. The coast of Europe were by and large "perfectly" lit, with no dark spots; no promising shipping lane was overlooked. Even the barren stretch of the Sahara whose banks had felled *la Méduse* boasted a first-class light by 1910. The ingenious design of Fresnel's lens made it, for the first time, easy to burn a light bright enough to satisfy all sailors. But if the technological story is simple, the historical one is not. Everywhere the lens popped up, behind it trailed the entangled questions of money, knowledge, and sea power that bedeviled nations everywhere. Fresnel's legacy, and its role in the unfolding modern age, proved his life to be much like the flash of light passing through his lens: short, perhaps, but brilliant enough to have a big impact.

ONE

DREAMS OF GLORY

AUGUSTIN FRESNEL was born on May 10, 1788, in Normandy, a year before the region ceased to exist. The celebrated Norman cow pastures, apple orchards, and windswept beaches were exactly where they had always been, of course, but as a political designation, "Normandy" disappeared when an early decree of the French Revolution replaced the traditional scheme of provinces with a new system of eighty-three departments. The change might have seemed simple on paper, but behind it lay the idea of forging one nation from a population fractured by different languages, customs, and identities. Before the Revolution, most people spoke a local dialect rather than French, and allegiances tilted more regionally than nationally. The architects of the Revolution, however, hoped that a new national spirit would be born with the new regime; they aimed to bind the distinct provinces together to create a single French culture. (Rewriting geography could only go so far: France eventually reintroduced "Normandy" as a regional designation.)

Developing a sense of national identity became crucial, as the new Republic soon found itself at war with the rest of Europe. It put on the field the first army ever assembled by national conscription, and motivated it by the novel means of patriotism. "La Marseillaise," adopted as the French national anthem in 1795, heralded the dawn of the new national senti-ment: "Allons enfants de la Patrie," the soldiers sang. "Le jour de gloire est arrivé!" (Arise, children of the Fatherland. The day of glory has arrived!) Fresnel himself was too young to heed the call, but as a member of the first postrevolutionary generation, he turned out to be very much a child of *la Patrie*. The notion of "glory" runs through his private writings and letters, fus-ing personal accomplishment with national ambition. Fresnel found its ultimate expression in his calling as a civil engineer,

Portrait of Augustin Fresnel.

one of a rising generation of technically trained professionals finding honor and prestige in the nineteenth century's massive state-sponsored public works projects. The engineering marvels of the century bound together the nation of France in a literal way, and testified to the glory of its genius and industry.

THE GENIUS OF THE ELDER BOUGHS

Fresnel's birthplace, Broglie, was one of the *ancien régime*'s fine aristocratic estates, home to the influential if aging Victor-François, duc de Broglie, who had served as marshal of France under Louis XV and Louis XVI. Fresnel's own parents were solidly bourgeois. His father, Jacques Fresnel, was an architect the duke had hired to renovate his chateau. His mother, Augustine, was the daughter of the chateau's keeper, François Mérimée, a noted lawyer. After marrying Jacques in 1785, Augustine gave birth to their first son, Louis, the next year, and then Augustin two years after that.

The estate of Louis XVI's former marshal was no place to be during the Revolution. The duke himself departed early on, ending up as an émigré in England. With the renovations to the chateau suspended, Jacques Fresnel found another job, and in 1790 he moved the family to Cherbourg, on the Normandy coast, to help build its massive new harbor, the only one between Dunkirk and Brest capable of sheltering warships. Augustin's earliest memories surely were rooted in this seaside town, the only time in his life when he lived directly on the coast.

As the Revolution intensified and France found itself embattled, construction in the harbor ceased. Jacques Fresnel, hoping to keep a low profile as the Reign of Terror unfolded, moved his family to his hometown of Mathieu, a small village outside of Caen where the Fresnels had lived for generations. The family property was decidedly modest, having been greatly reduced in the Revolution, but its quiet setting was now a welcome refuge from the turbulence. Augustin soon had two younger brothers, Léonor (born in 1790) and Fulgence (born in 1795).

Augustine taught her sons at home, and she made religious study the focus of their instruction. The household was strictly Jansenist, a movement begun by Cornelius Jansen in the 1630s and marked by asceticism and intellectual rigor. Throughout the seventeenth century, the rivals of Jansenism, most notably the rich and powerful Jesuits, claimed it was barely Catholic at all and almost indistinguishable from Calvinism (which it strongly resembled in both style and content). Jansenists, for their part, frowned at what Blaise Pascal, one of the movement's most prominent adherents, called the "moral casuistry" or laxity of the Jesuits. The movement nearly ended when Pope Alexander VII declared it heretical in 1655, but it saw a resurgence in the next century, as the Enlightenment's anti-clericalism targeted the Jesuits, culminating in their expulsion from France in 1764. The Jansenists, thus, in an "enemy-of-my-enemy" sort of way, were left free to occupy their niche as the somber, reverent Puritans of Catholicism. By the time Augustin was born, the movement had had a small but dedicated following among France's educated middle class.

Augustin remained a follower of the religion throughout his life. As an adult, he admired Pascal above all other writers, and even attempted to compose a few verses in his style. Augustin was not, however, a bookish child. Quiet and small for his age, and always in rather precarious health, he was, according to family legend, the slow one, lagging behind his brothers in their lessons. By the age of eight, he could barely read. As one account of his youth put it, "His memory seemed to refuse to learn words and his spirit, avid from childhood for positive experience, seemed to rebel against classical education." Although he struggled mightily learning Latin, he had an ingenious knack with less scholarly pursuits. He and his brother modeled an impressive array of medals from clay, carefully finding ways to heat them without causing breakage.

The children of the village of Mathieu perhaps saw something the adults missed: from age nine Augustin was known as a "genius" to his friends. He earned this title through the "experiments" he performed on the makeshift artillery they fashioned for their neighborhood battles. He tested all the local woods, both green and dry, to determine the most elastic and durable for the making of bows. He discovered the best length and bore for the elder-tree branches the children turned into canons for hurling projectiles. Augustin was so successful that the children's games, until then relatively innocuous, became downright dangerous, and the local parents united to ban them.

At thirteen, Augustin Fresnel left Mathieu to attend school in nearby Caen. The École centrale du Calvados, another Revolutionary experiment intended to make a nation out of France, was one of the central schools opened in each of the newly created *départements* to provide a uniform curriculum for secondary education. Fresnel, one of the youngest pupils, was sent only as company for his older brother, Louis, the most promising student of the family; the third son, Léonor, joined them the following year. Their timing was fortuitous: the science- and mathematics-heavy Écoles centrales, begun only in 1796, were shut down in 1803 and replaced with a network of lycées, which more strongly focused on the humanities.

As it was, the three brothers flourished. Louis, the star, won first prize in both grammatical logic and grammar, and praise as the "young man of the greatest worth, [who] unites talent, wisdom, and study." Léonor received the second prize in mathematics, with the note "the child gives great hopes." Augustin, for his part, took second place in Greek translation and earned the positive if less effusive comment "good subject overall like his two brothers."

In 1804, the year Napoléon crowned himself emperor, the brothers went on to the École polytechnique in Paris, the crown jewel of France's system of technical education. Founded during the Revolution to train engineers for the Revolutionary army, the school brought together France's top scientific minds and provided the most rigorous, up-to-date instruction of anywhere in the world. Students competed furiously to get one of the coveted spots, and it is a credit to Augustin's school in Caen that in the admission examinations he placed seventeenth out of the three hundred in his entering class. At sixteen, he was among the youngest of the students. Although the school recorded a rather unprepossessing description of Augustin— about five feet three inches tall, with chestnut hair, brown eyes, a small mouth, round chin, and oval face—and his work was mostly just fair, his skill at graphic arts and engineering drawing stood out, even among France's best and brightest, his dexterity and practical bent already marking him as an engineer.

After graduating in 1806, Augustin Fresnel continued his studies at the École nationale des ponts et chaussées (National School of Bridges and Highways), the world's oldest school of civil engineering. In 1808, the summer before he was to graduate from "les Ponts," he was assigned fieldwork building roads in the remote town of Périgueux, in the Dordogne. Arriving during the changeover of engineers, he found himself alone and immediately sought out the mathematics professor at the local school, M. Mellias. When the comte de Montalivet, general director of the Corps des ponts et chaussées (Corps of Bridges and Highways), wrote to the chief engineer at Périgueux to see how Fresnel was doing, his boss replied that Fresnel spent all of his free time with Professor Mellias, and that the two took long solitary walks together, apparently entertaining themselves by interrogating each other on the finer points of mathematics. Fresnel seemed to have no other acquaintances: "He does not appear to yet have the disposition for seeking out high society: his youth, his timidity, his awkward demeanor, . . . and the incorrect expressions which frequently escape from him, show that he has neither the taste nor the custom [for high society]." Fresnel was a nerd. Still, the chief engineer praised his skill. Time, he predicted, would surely make him a "good engineer and a kind man."

Fresnel graduated from les Ponts in 1809 with the title of *ingénieur ordinaire aspirant* (ordinary engineer in training). A dark event overshadowed the moment: his brother Louis, by then a lieutenant in the artillery, had died on the battlefield in Jaca, Spain, during an obscure, futile skirmish as Napoléon struggled to control the Iberian Peninsula. The loss left an enduring mark on Augustin: Louis and Léonor had been his two closest companions throughout his childhood and schooling. From now on, Léonor would be virtually his sole confidant.

Road Building and Rêveries

Fresnel's first assignment as an engineer sent him to the Vendée, then the most infamous region of France. Located on the west coast and far from Paris, the area was among the most isolated and undeveloped of the country, with few roads or even medium-sized towns; the average population of its communities was fifteen. When the Revolutionary government called for national conscription in 1793, requiring each region to supply 300,000 men for the army, the men of the Vendée refused and formed their own army to fight the Republican troops that came to enforce the national laws. The Vendée resistance, half counterrevolution, half civil war, flourished, mostly because of the region's inaccessibility. The initial success, however, led the central government to respond more fiercely, as national troops moved in to burn fields and forests, massacre livestock, and raze villages. By 1795, the Republican army claimed victory, at a staggering cost: out of a total population of 800,000, well over 100,000 Vendeans had died in the fighting.

When Napoléon took power in 1799, the Vendée was neither fully subdued nor integrated. Revolt, a perennial threat, often erupted into violence. In response, in 1804 Napoléon built a new town on the river Yon to serve as a military base in the region (naming it, with no false modesty, Napoléon-sur-Yon). Fresnel's particular task when he arrived in the Vendée was to connect this town to the surrounding area. New roads would end the isolation that had allowed the strong regional allegiances to grow, make it easier for Napoléon's soldiers to put down revolts, and serve as a model for the benefits of imperial rule. Fresnel was, literally, building the new nation of France. He hated his job, though, especially the tedious routine of overseeing men combing the vicinity for suitable rocks, breaking them down, and using them to level the road. "I find nothing more tiresome than having to manage other men, and I admit that I have no idea what I'm doing," he wrote home in despair. The work would be fine, he explained in another letter, if only his body wore out. Instead, it left his spirit "tormented by the worries of surveillance, and the need to reprimand and be the bad guy." Building a bridge over an irrigation canal or replacing a flooded dike provided a rare diversion, but his only real consolation came during the spare moments allowed him, when he would retire to his tent and "speculate on philosophical matters."

Fresnel's speculations often had a practical bent. He devised an improved design for a hydraulic ram. He invented a new process for cheaply making soda ash, an important chemical used to manufacture glass, textiles, soap, and paper. To attract attention to this process, Fresnel turned to Léonor Mérimée, his maternal uncle. Mérimée had moved to Paris as a young man to become an artist. He now studied painting preservation, and its chemical aspects brought him into occasional contact with France's leading scientists, including the prominent chemists Louis-Nicolas Vauquelin (who discovered beryllium) and Louis-Jacques Thénard (who discovered hydrogen peroxide). Vauquelin and Thénard looked at Fresnel's process but declined to pursue it, because, they claimed, it would not be cost-effective. Their loss: Fresnel's procedure was the same one, using ammonia, that allowed the Belgian Ernest Solvay to dominate the soda industry after 1861.

By all accounts, Fresnel was crushed at the lack of interest in his invention, but that did not impede his inquisitiveness. Sent to the south of France on a major project to build a road connecting Spain and northern Italy, he turned his attention to the subject of light. For his studies, he asked his brother to send him a physics textbook and to purchase for him a subscription to the journal *Annales de chimie* (*Annals of Chemistry*). Fresnel's professors at the École polytechnique had been strong adherents of Isaac Newton's theory describing light as tiny, weightless particles. Fresnel now focused on a phenomenon that had always troubled Newton's system: diffraction. By this strange trick of light, the shadow produced by a knife, razor, or other thin edge was inexplicably fringed, with bands of light and darkness alternating in patterns that varied with the object. Stranger still, when a beam of light passed through a small slit, it seemed to break apart and spread across the room, producing a pattern of light and dark much wider than either the original beam of light or the slit through which it passed. The effect had been known since the seventeenth century (its name comes from the Latin *diffringere*, "to break into pieces"), but no one had been able to fully explain it. Newton himself had spent an inordinate amount of time investigating the phenomenon, arguing that there was some kind of force interacting between the light particles and the edge. But his explanation was unsatisfying: how could a light particle be there one minute and not the next?

Fresnel believed he could clear up the issue by considering light as a wave rather than as a particle. It was hard to imagine how combining light particles could add up to

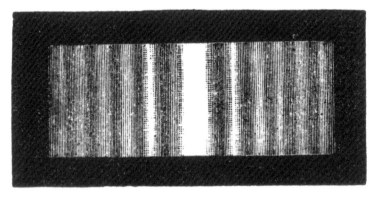

The diffraction pattern produced by shining a light through a thin slit. Although the width of the slit is the same as the white band in the middle, additional bands of light spread out on each side in patterns of alternating dark and light fringes.

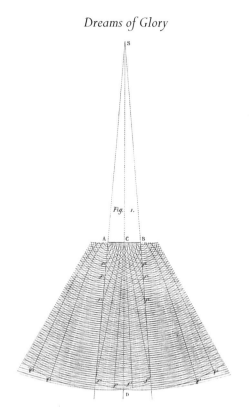

*Fresnel's illustration of the propagation of two waves
coming through parallel double slits. They combined, on the screen,
to form alternating bands of light and dark.*

darkness. But if two waves lined up with one another, so that
the peaks of one wave matched precisely the troughs of the
other, then combining them caused the wave to disappear
entirely, as each peak and trough canceled each other out (a
process known as destructive interference). If, on the other
hand, the two lined up so that the peaks of one matched the
peaks of the other, combining them would yield a single wave
whose peaks were now twice as tall (known as constructive
interference). Fresnel created an elegant account describing

the effect of this interference: the bright fringes were where light waves interfered constructively, and the dark ones were where they interfered destructively.

Fresnel wrote up his ideas in an essay he called his "*rêver-ies*," or daydreams. He sent it off to his uncle Mérimée, who passed it on to André-Marie Ampère, an up-and-coming physicist who would soon make a name for himself laying the early foundations of electrodynamics. But to Fresnel's dismay, Ampère "lost" the essay, unread.

What followed was a stroke of luck. While attending a fashionable dinner in Paris, Mérimée found himself seated next to François Arago, a lively young physicist who was prob-ably the only man in Paris both capable of understanding and interested in supporting Fresnel's unorthodox theory. Two years older than Fresnel, Arago had been just one year ahead of him at the École polytechnique, though they had traveled in very different circles. Even as a student, Arago belonged to the rarefied upper echelon. He cultivated relationships with the scientific elite, and his contacts paid off in 1806, when the Bureau des longitudes (Bureau of Longitudes) chose him, at age twenty and still officially listed as a student, for one of its most important and high-profile scientific expedi-tions: extending the measurement of the meridian, the arc of the earth running from the North to the South Pole. The Académie des sciences (French Academy of Sciences) had defined the meter as one ten-millionth of the meridian in 1791, and had sent out two men to measure a portion of its arc through the painstaking process of triangulation. The men climbed hilltops and church towers to shine light signals at each other from these locations, and after carefully measur-ing the angles and one side of the triangle, they calculated the remaining distances. They repeated this process dozens of times south from Paris until one of the men met an untimely death in Spain. Arago, along with his friend and fellow physicist Jean-Baptiste Biot, stepped in to continue as far south as they could, which turned out to be the island of Ibiza, off the coast of Spain. With most of the measurements done in 1808, Biot returned home while Arago stayed for a few final tests. In these intervening weeks Napoléon's forces (including the unfortunate Louis Fresnel) invaded Spain, and the local Ibizans, deciding that the lights Arago had been shining as part of his experiments looked suspiciously like spy signals, threw him in jail. The daring Arago escaped (twice), only to have the ship he was on captured by pirates as it approached the French coast. He ended up in North Africa, where he walked through the desert disguised as an Arab and became a slave of the dey of Algiers. Meanwhile, the Bureau des longitudes gave him up for dead, ended his stipend, and sent their regrets to his parents. By the time he made it back home, it had been well over a year since anyone in France had heard from him.

Arago's dramatic return from the dead in 1809 became the talk of the town. An exceptional raconteur, he thrilled audiences across the salons of Paris with his harrowing tales of intrigue and danger, all in the name of science and France. At that moment, the Académie des sciences had an opening for a new member. Although election to the Académie represented the highest scientific honor in France, one usually reserved for venerable statesmen, Arago's sudden celebrity placed him among the top candidates. The one major obstacle to his selection was the unyielding opposition of France's greatest physicist.

Pierre-Simon Laplace was the most powerful man in French science, praised for his focus and determination more than his charm. His thorough and understated biographer,

Charles Coulston Gillispie, wrote that "not a single testimo-
nial bespeaking congeniality survives" in the historical record.
Laplace's fortunes had risen with those of Napoléon, a former student who fancied himself as a bit of a physicist. Upon
assuming power, Napoléon appointed the slender Laplace to
be minister of the interior, an unfortunate choice that lasted
scarcely six weeks; Napoléon's rather exasperated memory of
the affair was that Laplace reveled in unnecessary complica-
tions and "in short carried the spirit of the infinitesimal into
administration." Nevertheless, Laplace enjoyed a prestigious
and lucrative (if largely symbolic) career as chancellor of the
Senate throughout Napoléon's reign.

By 1805, Laplace had completed the last volume of his
Traité de mécanique céleste, or *Celestial Mechanics*, a magisterial
work that finally solved the most pressing scientific problem
of the past century: the instability of the solar system. Newton's account of planetary motion had solved the straightfor-
ward case of how a single planet revolves around the sun, but
the mathematics became intractable as soon as another body
was added. It seemed that adding an additional planet would
pull the first planet slightly out of its orbit, the delicate balance of forces would be disturbed, and the planet would go
spiraling into the sun or out into the cosmos. Newton saw
this as welcome proof that God was directly intervening in
the solar system to keep everything in equilibrium, but few
shared his view. Physicists thus considered it a triumph when
Laplace was able to demonstrate, mathematically, that an
additional body would cause a planet to oscillate between a
maximum and minimum, but it would remain stable. It was
this proof that the planetary system could run on its own
that underpinned the famous moment when Napoléon, upon
receiving the final volume of Laplace's *Traité*, reportedly

asked the physicist where God fit into the heavenly scheme, and Laplace replied, "Sire, I have no need of that hypothesis."

Laplace then turned his attention to the more earthbound physical forces—light, heat, electricity, and magnetism—to give them the same treatment Newton had given celestial bodies. The key to Newton's work was formulating gravity as a force that got weaker as the distance between two objects increased, with the magnitude inversely proportional to the square of the distance. Laplace, accordingly, sought to explain light through a similar inverse-square force acting on particles of light, and he gave several students, including Arago and Biot, the task of exploring optical phenomena within that framework. The particle theory of light thus underlay his broader vision for a triumphant Newtonian worldview, making Fresnel's work on diffraction deeply heretical. By 1809, however, Arago had begun to chafe at the reins of his former mentor.

Arago actually disliked Laplace the very first time they met, and often recounted the story: As a young student at the École polytechnique, Arago accompanied his friend Siméon-Denis Poisson to dine with Laplace at his home on the rue de Tournon. Arago was thrilled to meet the eminent scientist, he recalled, "but what was my disenchantment, when one day I heard Madame de Laplace, approaching her husband, say to him, 'Will you entrust to me the key of the sugar?'" Laplace was a bully, and often a petty one at that. The man who understood the heavens better than anyone else locked the sugar away from his wife.

Laplace's opposition to Arago's candidacy at the Académie des sciences cemented the young man's disillusionment. Laplace had his own candidate—the very mathematician Poisson who had introduced the two—and worked tirelessly to undermine Arago. In an unusual breach of protocol the day before the ballot, Laplace, Arago said, "was active and incessant to have my admission postponed." When the election was a landslide victory for Arago, winning forty-seven of the fifty-two votes cast, Laplace leaned heavily on Arago to refuse the nomination until there was a place open for Poisson.

With the rupture between the two physicists complete, Arago, quick to denounce tyranny in any form, took aim at Laplace's stranglehold on French physics, as well as the explanatory framework of particles and forces he was trying to impose on the discipline. Arago's next paper on optics, published in 1811, intimated that the Newtonian system might not be entirely correct, but he did not elaborate and was in all likelihood still searching for a route of assault on the Laplacian view. So when Mérimée mentioned his nephew's work during that fortuitous dinner in Paris, Arago saw just the opportunity he needed. Fresnel, and his renegade theory of light waves, became his new project.

The two men were a study in contrasts. Accounts of Arago rarely failed to mention his "tall commanding form, . . . his full sonorous voice, his striking features, and dark piercing eyes, shaded by thick bushy brows." Fresnel, on the other hand, more often was described as "pale" and "delicate." A seemingly boundless energy accompanied Arago's robust frame. He was charming, lively, a bit hot-headed, but immensely popular: everything, in short, Fresnel was not. The two could almost be thought of as hailing from different countries. Arago grew up in Perpignan, at the foot of the Pyrénées, in the Catalan region. (His family name even derived from the Kingdom of Aragon, which spilled over from Spain.) French Catalonia, like Fresnel's Normandy, maintained a strong local identity up until the nineteenth century. Arago, who spoke Catalan, played up what he called his "*ardeurs méridionales*" (southern passions). With Fresnel as the picture of northern phlegmatic self-possession, both seemed determined to live up to their regional stereotypes. Religion, politics, class: all the issues that were supposed to matter in the nineteenth century divided them. One was a cheerful agnostic, the other a stern Jansenist; one a left-wing radical, the other a royal loyalist. One had seen his family's fortune flourish under the Revolution, the other had been impoverished by it. A more unlikely pair simply could not have existed, but their careers soon became inseparable.

Portrait of Francois Arago.

FRESNEL'S HUNDRED DAYS

Fresnel, innocent of Parisian politics, had no idea how inflammatory his suggestion of light as waves would become, or how viciously Laplace would oppose it. But his timing was fortunate in a period marked by dramatic reverses. When Napoléon fell from power in 1814, defeated by the combined powers of Great Britain, Prussia, Austria, Sweden, Russia, Spain, and Portugal, Fresnel, the lifelong royalist, enthusiastically greeted the return of Louis XVIII from exile and his crowning as king of France. Napoléon, however, escaped from his exile on Elba in February of 1815, landing in the south of France with the intention of marching on Paris. Fresnel was stationed in the southern town of Nyons, virtually on Napoléon's path, and without leave, he immediately joined the army being raised nearby by the duc d'Angoulême, the king's nephew, to cut Napoléon off. The effort failed spectacularly, and Fresnel returned, chastened, to Nyons.

With Napoléon back in power, Fresnel's actions looked downright treasonous. Denounced to the local police, accused of "show[ing] himself constantly to be the enemy of the Government," he was greeted by boos and catcalls in the streets. A military order suspended him from his position and placed him under surveillance. Fresnel's boss at Nyons, chief engineer Pierre Charles Le Sage, had a hard time believing this shy, scrupulous engineer could be an enemy of the state. He immediately wrote to comte Louis-Mathieu Molé, the new, young director of the Ponts et chaussées. "When one knows Mr. Fresnel," he pointed out, "it is difficult to believe the accusations against him." Fresnel was, he said, "very measured in everything he says" and "un homme froid" (literally a cold man, likely meaning cool and self-possessed). He added that Fresnel was a skilled engineer who would be sorely missed.

Fresnel was allowed to return to his hometown of Mathieu to serve his suspension. Stopping in Paris during his trip north, he met with Arago for the first time. The walk from his uncle's house on the place de l'Estrapade, where he was staying, to Arago's residence in the Observatoire de Paris (Paris Observatory) was a short one, so they saw each other frequently. Fresnel found Arago enthusiastic and encouraging. Arago provided him with a list of the latest works on optics, including one by the English physician Thomas Young, who had been working on his own wave theory of light. Fresnel never made it through the text—the English was indecipherable to him—but he left Paris with a renewed conviction of the importance of his studies on diffraction.

In Mathieu, he set to work on his experiments. Almost as soon as he arrived, however, Napoléon's sound defeat at Waterloo changed the situation once again. The emperor's return from exile had lasted only a hundred days. Now Fresnel was a hero, receiving a certificate of praise for his devotion to the Crown. Molé ordered him reinstated, although Le Sage hinted to the director that sending Fresnel back to Nyons, "the residence in which had seemed to displease him," and where he had been so recently persecuted, would not come as much of a reward to him. Indeed, as soon as Fresnel learned that he had been recalled to his post, he asked for a two-month leave of absence because "the poor state of my health does not permit me to return immediately to Nyons," a request that was easily granted since road-building operations had been suspended.

Fresnel intended to provide the most precise experimental account of diffraction ever attempted, which he would then compare to the theoretical results derived from his wave theory. With help from the village locksmith, he set up a dark room, completely blocked off from the sun except for a narrow beam of light, which came through a hole he punched through a metal sheet on the outer wall. To get the beam as bright as possible, Fresnel used a large lens on the outside to focus the sun's light on the hole. Unfortunately, the sun's motion across the sky quickly displaced the focal point, and Fresnel found he had only a brief instant to make his observations. Arago had suggested using an extremely convex lens to mitigate the effects of the sun's motion, but such specialized equipment was unavailable in Mathieu. Instead, Fresnel improvised, placing a drop of honey, from the bees his mother kept, on the hole in the metal sheet. He noted proudly the excellent result his homemade solution provided.

Inside the room, he then placed a metal wire (the first of several test objects) in front of the light source, so it would cast its shadow on a white screen set on the wall opposite to where the sun shone in. Lacking a micrometer, the standard device scientists used to measure small distances, Fresnel first used a ruler to measure the distances between the shadow's fringes. But the measurements were not as precise as he wanted. He found that when he looked directly into the beam of light, removing the screen and placing his eye where the shadow from the wire would fall, he could see the fringes better. To measure them, he rigged a triangular grid of silk thread, which he placed between his eyes and the wire. After noting where on the triangle the light and dark bands fell, he used geometry to compute the width of the fringes.

His observations matched his calculated predictions to .2 millimeter, "leaving no doubt," he wrote, about the correctness of the formula derived from the wave theory. Excited, he wrote up his results in the fall of 1815 and sent the memoir to Mérimée. As the Académie des sciences considered only works submitted (although not necessarily written) by its members, Mérimée then forwarded it to Arago, their only champion on the inside. In November, Arago sent Fresnel the news that the Académie had selected Arago to review it. (The Académie as a whole never heard more than the reviewer's account of a paper.)

Fresnel's leave from his engineering work expired, and he was assigned to yet another road-building job, this time in Rennes, in the northwest corner of France. He begged Molé for another month of leave, or at least an appointment that would allow him to remain in Normandy. "My health has been weakened by the excessive heat of the Midi and needs a colder climate to restore itself entirely. It would also be sweet for me to spend a few years near my mother, from whom I have been separated for so long. It would only be with greatest regret and worry that I would leave her at a moment when the Prussians are inundating Normandy." Moreover, he complained, he had "done nothing but build roads," and he wanted to undertake the "construction of works of rather considerable art." This cavalcade of excuses omitted the one truly pressing reason: Fresnel was on the verge of revolutionizing the science of light.

Now he had an important ally. Arago approached Gaspard de Prony, director of Fresnel's engineering school, whom he knew from both the École polytechnique and the Académie. A noted mathematician, Prony had done much to ensure the rigor of the engineering education at les Ponts. Fresnel's experiments, Arago told Prony, were "very remarkable for their novelty," and Arago wanted to see them for himself. Prony, in turn, wrote to Molé the next day, supporting Arago's request to have Fresnel come to Paris. Fresnel's absence from his job, Prony stressed, "will be profitable both to the progress of science and the glory of the Corps of Roads and Bridges." Molé accordingly granted Fresnel one month's leave to repeat in Paris the "very remarkable experiments" he had heard so much about.

Fresnel arrived in the spring of 1816, just in time to catch a long stretch of sunless days. Even when he gained an extension of his leave through the summer, the clouds and rain continued, with no end in sight. Fresnel's time in Paris coincided with the infamous "year without a summer," known for its meteorological abnormalities. Now thought to have been caused by a massive volcanic eruption in Indonesia the year before which spread a cloud of ash around the world, the strange weather was marked by a persistently hazy sky and abnormally cold temperatures in the Northern Hemisphere. Snow lined the streets of New York in June. Water buffalo froze in China. At Lake Geneva, the weather kept Lord Byron and his houseguests indoors, where they told ghost stories and created the legends of Dracula and Frankenstein. Fresnel, meanwhile, waited with frustration for days bright enough to run his experiments.

Fresnel wrote home that Arago "attached the greatest importance to my discovery" and pressed Fresnel to write up the results so that Arago could publish them in *Annales de chimie et physique*, a journal he had configured as a vehicle for his campaign against Laplace after he, Arago, took over editorship of *Annales de chimie* in 1815. Fresnel completed the article and sent a copy to Molé, who replied, "You could not give your leisure a more noble goal," praising the honor it brought to the Corps des ponts et chaussées.

Nevertheless, disappointment was in store. Arago returned from a trip to England, where he had talked up Fresnel and his discoveries to no avail, as the English assumed Fresnel was simply repeating the work of Thomas Young. (Young had not, in fact, achieved the close quantitative agreement between experimental and theoretical results that Fresnel had.) Fresnel wrote dispiritedly to his brother, "I have decided to remain a modest engineer of bridges and roads, and even to abandon physics, if circumstances require it. I would resolve to do so the more easily that I now see it's a stupid plan troubling oneself to acquire a small bit of glory, that they'll always quarrel with you about it. . . . Fie on contested glory!"

The dismal summer weather led to crop failure across the globe. Brittany, still Fresnel's official post, was particularly hard hit: the streets of Rennes filled with men, women, and children, "all pale with hunger," starved out of the barren countryside. As the famine reached catastrophic levels, the French administration set up a number of *ateliers de charité* (charity workhouses), where the indigent could make a living by helping out with public works projects. In October, Molé recalled Fresnel from Paris to oversee those in his district, and he spent the winter supervising both his regular road crew and the charity laborers, 150 men in all. He hated every minute of it. "I am a third of the time on horseback," he complained, as he had to constantly make the rounds to supervise the many projects spread throughout his jurisdiction. The heavy administrative load, the number of sites to oversee, and the traveling left him exhausted, with no time for his own research. "I burn to leave Rennes," he told Arago, and at winter's end, when plans to dismantle the workhouses were under way, Fresnel once again began petitioning Molé for more time off or a position closer to Paris, reminding him of his own words about the honor of the Corps des ponts et chaussées. Fresnel said he was "impatient" to return to his experiments; a delay of even a few weeks might cause him to lose "the advantage of priority." Molé remained unmoved.

Then, in March of 1817, the Académie des sciences announced its annual prize competition for what it deemed the most pressing scientific problem of the day: diffraction. A sizable monetary reward and ample prestige would accompany the award. Fresnel again asked for a leave of absence to compete, and now his request was accompanied by support from Prony, confirming the importance of the research. Since Molé had recently been made minister of the navy, his successor at the Ponts et chaussées, Louis Becquey, would decide Fresnel's fate. Significantly older and less personally ambitious than Molé, the new director planned to extend the scope of France's public works, particularly through the building of canals.

Becquey, as Arago put it, "on every occasion treated Fresnel with the kindness of a father," and granted him a few months leave to work on his prize submission. By the autumn of 1817, Fresnel was headed back to Paris. Using a powerful mathematical technique he had invented for wave fronts (or the front lines of waves as they move forward), now known as the Fresnel integrals, he calculated precisely how the light waves would propagate into particular diffraction patterns for differently shaped edges, and then compared his predictions to his experimental results. He handed in his memoir on April 20, 1818. Days later, he returned to his engineering work, starting in a new position on the canal de l'Ourcq that Becquey had arranged for him. The sixty-seven-mile-long canal was Becquey's first major

project at Ponts et chaussées, a long-overdue attempt to provide Paris with decent drinking water from the river Ourcq,
since the Seine had grown unimaginably polluted. It was not
a permanent position for Fresnel, but it would keep him near
Paris as the Académie evaluated his memoir.

In the war between corpuscularians and undulationists
(as supporters of the particle theory and and those of the
wave theory of light came to be known), the Académie des
sciences made no secret of its sympathy for particles. It
had just awarded a medal to David Brewster, Britain's most
ardent defender of the theory, claiming his optical work the
most important advance in physical science over the past
two years. Corpuscularians dominated the prize committee evaluating the work on diffraction. Laplace himself was
its senior member, with two of his closest acolytes, Biot
and Poisson, joining him. Arago's friendship with Poisson
had evaporated during the fight over the seat at the Académie des sciences. Biot, meanwhile, had gone from Arago's
closest collaborator to his most bitter rival, as he horned
in on Arago's work, represented Arago's ideas as his own,
and attacked Arago's claims while defending the Newtonian approach. It was Biot who had suggested the topic of
diffraction for the prize competition, hoping to finally find
a satisfying corpuscularian explanation for the troublesome phenomenon. Even the language of the competition's
announcement was couched in terms that assumed light was
a particle; it asked entrants to infer "the motions of the rays
in their passage close to bodies."

The deck was clearly stacked against Fresnel. According
to Mérimée, Arago decried the competition as an "enemy
attack," warned the members of the Académie of Laplace's
clear "wish to dominate," and complained of "the impropriety

of nominating party men." Arago's efforts gained him inclusion on the prize committee, as well as a fifth member, the chemist Joseph Gay-Lussac, who was generally considered impartial. Arago had won this battle for the undulationists— the war, however, was far from over.

The contestants submitted their entries anonymously, but Fresnel's was easy to identify. Tagged with the epigraph, "Natura simplex et fecunda" (Nature simple and fertile), it presented the sole defense of the wave position. Fresnel had

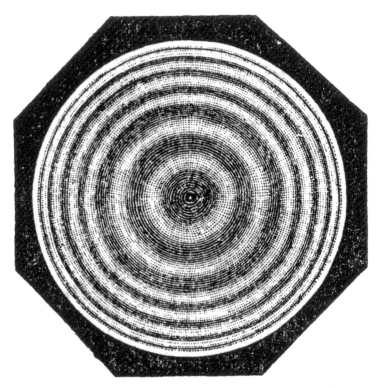

The surprising shadow of a circular disc. As Arago verified (though not himself expecting it), at the very center of the shadow was a bright, white spot.

worked through twenty-five possible cases where diffraction patterns could be seen in an object's shadow, from thin wires to razors' edges. In each case he calculated the expected intensity pattern of the shadows, and then compared his predictions to his experimental results. The agreement was striking. Nevertheless, the corpuscularians did not accept it, and Poisson found what he took to be a fatal flaw in Fresnel's work. Using Fresnel's equations, he had worked through an example that Fresnel himself had not tried, the shadow cast by a circular disc. The calculations predicted what could only be a physical absurdity: that at the very center of the disc's shadow, where one would expect it to be the darkest, there would be a bright, white spot, as if the disc was not there at all. The corpuscularians reveled in the idea that any theory yielding such a patently implausible result must be wrong. But the anomaly merely spurred Arago to action. He quickly pulled together, with the help of instrument maker François Soleil, a test of the diffraction pattern of a disc. The entire prize commission assembled to watch—and there in the center of the shadow was the white spot in question, "just as illuminated as if the screen did not exist." Fresnel's improbable calculations had been right.

The Académie awarded Fresnel the Grand Prize on March 15, 1819. Laplace, Biot, and Poisson would never accept the wave theory of light, but there was no denying the promise of Fresnel's work. His years of toiling in the provinces, begging for leave, were over. Arago now asked Becquey to transfer Fresnel to Paris to help him with work he had just proposed for the Commission des phares (Lighthouse Commission). The motivation, no doubt, was to have his new best ally close at hand. Lighthouses, though, proved an inspired vocation for the quiet engineer. Little did he know his days of bringing glory to France were just beginning.

TWO

THE FLASH OF BRILLIANCE

NAPOLÉON CREATED the Commission des phares on April 29, 1811, at the height of his empire, which included all of continental Europe except the Balkans. His vision of the Commission was appropriately grand and brimming with imperial ambition. The decree promised to improve the lighthouses "on the universality of the coasts of the Empire," a twenty-four-thousand-mile shoreline that stretched from the Baltic Sea to the Straits of Gibraltar and around to cover the Tyrrhenian Sea and the western edge of the Adriatic. The lighthouses were perhaps a military necessity, but more than that they symbolized the engineering prowess of enlightened France. Central to Napoléon's imperial logic was the vision of himself as not merely a *conqueror*, but an agent of civilization. He would bestow upon his subjects the benefits of progress and modernity. What better symbol than an actual beacon of light? Practical, powerful, and grounded on the principles of science, lighthouses combined several of Napoléon's favorite strands of culture. His own training in the artillery had emphasized mathematics, and he strongly supported a rigorous technical training for engineers. He even had himself elected to the Académie des sciences, which he attended enthusiastically. Unsurprisingly, then, he decided to put control of the lighthouses, formerly the responsibility of the navy, in the hands of engineers. The new Commission was now under the elite Corps des ponts et chaussées, and specifically the young comte de Molé.

Molé immediately assembled an impressive roster of high-profile names to fill the nine-member commission, with three men from the Ponts et chaussées, three from the navy, and three from the Académie. Joseph-Mathieu Sganzin, inspector general of the Ponts et chaussées, headed the Commission. With the other two appointed engineers—Pierre Ferregeau, director of the Travaux Maritime (Maritime Works), and Jean Bernard Tarbé de Vauxclairs, who had built several defensive ports—Sganzin often accompanied Napoléon on tours of his newly acquired coastline. The men from the navy—prominent ship captains Joseph-Saturnin Peytes de Montcabrié, Louis-Léon Jacob, and Louis-Isidore Duperry—were occupied at sea, and during peacetime they were required to reside in the maritime *département* assigned to them. With these six members spending much of their time outside Paris, convening a meeting of the Commission proved difficult.

The three Académie scientists generally stayed in Paris, but lighthouses were not their foremost concern. Jacques Charles, known for Charles's law, relating the volume of a gas to its temperature, was celebrated in his time as the inventor, along with the Montgolfier brothers, of the hot-air balloon and as the first man to ride in one, in 1783. When the Commission was established, Charles, age sixty-five, was fully occupied with teaching duties. Jacques-Noël Sané, a wizard of ship design, and in his seventies, showed little interest in the Commission. Étienne-Louis Malus had recently presented startling new discoveries on the polarization of light while still in his thirties, but in the end he could do little for the Commission. Having accompanied Napoléon's expedition to Egypt in 1798, Malus had been placed in charge of a makeshift hospital tending to the hundreds of men who fell ill with the plague. He had caught the disease but seemed to fully recover. More than a decade later, and only weeks after his appointment to the Commission des phares, he fell ill again and died within the year.

The energetic and ambitious Arago replaced Malus on the Commission in 1813. But Arago, like the others, had little spare time. In addition to his duties at the Académie, he was chair of analytical geometry at the École polytechnique and secretary of the Bureau des longitudes. He also served as the de facto director of observations at the Observatoire de Paris, overseeing the instruments and other astronomers, and giving an enormously popular series of public lectures. He continued his own research in optics, and had a new wife and a social schedule that alone would have exhausted most others.

The Commission had not met at all by the time Napoléon's empire began to crumble, and by and large it continued to not meet for the next several years. The group largely contented itself with the upkeep of existing stations. It did debate a system of color-coded lights intended to distinguish the various stations from one another, but the colored lights were necessarily much dimmer and harder to see, so the Commission scrapped that project. Little had been done by 1817, when comte de Molé was promoted to director of the Ministère de la marine (Ministry of marine) and replaced at the Ponts et chaussées by Louis Becquey. There followed a blistering report from Molé's Ministry that detailed a long list of sailors' complaints and pointed out the inferiority of French lights to their English counterparts.

Becquey, whose responsibilities included lighthouses, had to do something. His first idea was a national contest, with a sizable reward for anyone who could improve France's maritime navigation. Arago, however, persuaded his fellow

Commission members to let him run a series of experiments on the existing technology.

The equipment in question was relatively new. Since antiquity, illumination had been provided by flames, first from an open wood fire, then from coal fires in the seventeenth century and candles in the eighteenth. Only in the 1780s did two new developments render the arrangements more scientific. A smokeless lamp, designed by Swiss physicist Aimé Argand in 1780, soon made oil the most popular fuel for light throughout Europe. Around the same time, lighthouse builders began placing mirrors behind the lamps to reflect as much light outward as possible. Although at first simply spherical bowls, the mirrors quickly took on the more efficient shape of a parabola, earning them the name "parabolic reflectors." To increase the light, reflectors could be stacked on top of one another. Several of them could be arranged around a device that rotated slowly, so that the lamps blinked in and out of visibility. The characteristic flash of each rotating apparatus, as it was known, allowed sailors to determine precisely where they were on the coastline.

Two men supplied all of France's reflectors at the time. Étienne Lenoir, an instrument maker who specialized in sextants and surveying equipment, made the first reflecting lighthouse light for Louis XVI in 1788. Consisting of twelve parabolic mirrors, each two and a half feet wide, it was initially displayed at the end of the Grand Canal de Versailles, then dismantled and shipped to the coast, for use in the Cordouan lighthouse. Lenoir went on to furnish reflectors for several other lighthouses, at Calais, Le Havre, and Saint-Mathieu. The second supplier, Isaac-Ami Bordier-Marcet, a "lampist-engineer" who also provided gas lighting for city streets and illumination for churches, began constructing the reflectors soon after Lenoir did, furnishing the lighthouses at Four, Cap Fréhel, and

Barfleur. Yet the adoption of the lamp-and-mirror combination, although widespread, resulted in only incremental improvements, and complaints of inadequate illumination continued.

Arago knew well how disappointingly feeble the light from the reflectors was: he had, when he was measuring the meridian with fellow physicist Jean-Baptiste Biot, spent months on a mountaintop searching for any sign of light from one even though they had, in their desperation, used up to eight mirrors to generate a single beam, an option that would not be possible for the rotating lighthouse devices. Arago proposed to run a series of tests on French and British reflectors. But, he told Becquey, he needed two assistants: Claude-Louis Mathieu, a close friend and colleague at the Observatoire (and soon-to-be brother-in-law), and Augustin Fresnel.

Becquey consented, although he emphasized that the appointment was not going to relieve Fresnel of road duty. He wrote to the Commission:

> 21 June 1819
>
> Sirs, I have the honor of informing you that, following the desire expressed in your report, I have decided that M. Fresnel will be *temporarily* placed at your disposition. I announce to this engineer that he should assist you in your work in the intervals which his work on the Paris roads allows.
>
> I have the honor, etc., Becquey

Arago outlined several tests designed to compare France's best reflectors, including the "grand Lenoir" at the Cordouan lighthouse and the Bordier-Marcet reflector, with a parabolic reflector brought over from England, made by George Robinson, supplier of the English lights. Most of the work was done at the Observatoire, where both Arago and Mathieu lived and had

The central principle behind the parabolic mirror. Under normal conditions, the light of a candle will travel evenly in every direction. A curved mirror placed behind the candle will reflect the light rays so that they all head in the same direction.

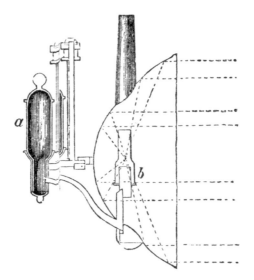

The lamp had to be carefully formed in the shape of a parabola to ensure the reflected rays formed a parallel beam. In the completed reflector, the flame from the oil lamp is in the center, with the reservoir of oil behind it.

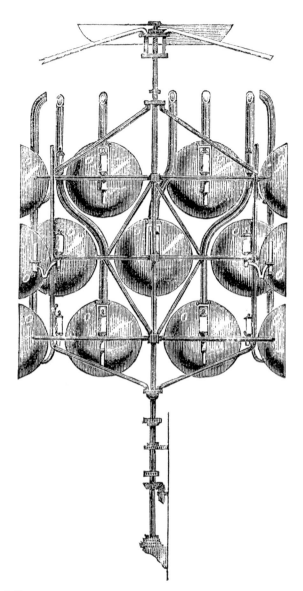

A fully assembled revolving apparatus would consist of several reflectors, facing outward in a circle so that their beams shine out successively as the device rotates, creating a flashing effect.

access to optical equipment. The big test, however, involved mounting the lights on the top of the Arc de Triomphe so the men could observe them from a distance. The famous Arc at the place de l'Étoile was still under construction, but nonetheless it provided the best high point around. Fresnel, the designated engineer of the group, found himself continually climbing up to secure the equipment. He complained to his brother that he would have preferred to work on only the optical part of the testing, but his responsibilities included dealing with "the hassle of these little details." Mounting the reflectors was a delicate process, since so much depended on the alignment, and Fresnel soon learned firsthand how difficult it was to coax light out of them. While the tests on the reflectors were being conducted, Fresnel started to search for a solution to the problem.

FRESNEL INVENTS THE LENS

Fresnel "perceived at first glance where the difficulty lay," as Arago put it. Even the most flawless mirror lost half its light on reflection (half is absorbed, half is reflected), and in practice, it was inevitably worse, as the mirror was never perfectly reflective, the parabolic geometry was hard to get right, and the device needed a hole for the lamp's burner to fit through.

All could be solved, Fresnel realized, by replacing the reflectors with lenses. A lens, by refraction, or the bending of light, could do the same thing that the mirrors did by reflection: that is, it could take all the light emanating from the source and direct it into a beam. Where as mirrors automatically lost half their light a lens of reasonable thickness would give up only one-twentieth or so. But this setup had a fatal flaw. Because a lens would have to be placed close to the light source to catch as much of the light as possible, the light rays would have to be refracted through a very large angle to get them parallel with one another, in order to create a bright light. The short focal length needed to bend light sharply required a steep curvature, a lens with a middle much thicker than its edges. The thicker the middle, however, the the greater the loss of light traversing it. The English had installed such a lens at the Portland Bill Lighthouse in 1789, with abysmal results: five and

Section of Lens

The Portland Bill Lighthouse. Several biconvex lenses (one is shown in profile to the right) were installed around the sides of the lantern room.

a half inches thick at the center, the lens lost far more light than any reflector. Considered a failure, its use never caught on.

A larger lens could be placed farther from the light source, and thus reduce the curvature required. Yet, Fresnel realized, such an object would have to be enormous, with a diameter larger than any high-quality lens yet made. He briefly considered solving the problem by filling two convex pieces of glass with (ever the Frenchman) distilled wine: the clear liquid could refract the beam and lose much less light. Unfortunately other problems presented, such as leakage and the formation of films. In any cases, a giant lens, whether solid glass or filled with wine, would be too heavy for any of the existing machinery to rotate it regularly in an arc.

Fresnel then thought of a way to create a giant lens without all the bulk in the middle: construct it in steps. Break the single curved surface into several concentric sections consisting of distinct prisms, or triangular pieces of glass that would refract the light. Each of these individual prisms would bend the light rays from their source into a parallel line. Added together, the prisms would send this light out in a single beam, precisely as one big lens would. The system was now trim and lean. Fresnel called it his "lenses by steps," or *lentilles à échelons*. It would form the foundation of a system based on refraction called dioptric (from the Greek *dioptrikos*, a surveying instrument) and improve on the deficiencies of the mirror system, based on reflection, called catoptric (from the Greek word for mirror).

Fresnel mentioned his idea in his first meeting with the Commission des phares in August of 1819, after he presented his official report on the tests of the reflectors (all of which he found deficient). Sganzin, head of the Commission, admitted that he did not quite understand the concept. Charles, the ballooning chemist, recalled that he had read something similar in the works of Georges-Louis Leclerc, comte de Buffon, an Enlightenment *philosophe* full of wide-ranging and often fanciful ideas about how nature worked. Fresnel was crushed to learn that he had once again "broken through an open door," as he had done earlier with his wave theory of light. But there were several crucial differences between his plan and Buffon's. Buffon had been more interested in concentrating the effects of heat than light, and had been hoping for something more like the "burning glasses" that, legend had it, Archimedes had used to focus the sun's rays on attacking ships to set them aflame. Buffon's design was for a single piece of glass with steps cut out from it, and he had produced nothing more than small, handheld models. Fresnel insisted that separate prisms made the operation possible.

To persuade the Commission, Fresnel realized he had to construct a lens that worked, and it would not be an easy task. He decided to use a plano-convex shape (flat on one side, curved on the other) instead of Buffon's biconvex form (curved on both sides). The first step was to find suitable glass. Here Fresnel had some tough choices to make. Flint glass was usually the preference for precision optics. Also known as leaded crystal because of its high lead content, flint glass was prized for its brilliance, clarity, and high refractive index,

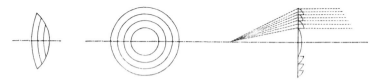

*Fresnel submitted this sketch to the Commission des phares.
It shows a thick convex lens on the left. Fresnel proposed to cut away
sections in four steps, which would produce the thinner lens on the
right, still capable of bending the light into a single beam.*

which meant that it bent light more sharply than any other kind of glass. Flint glass, however, was roughly twice as dense as most other types of glass and particularly heavy, as anyone who has toasted with a lead crystal goblet knows. For these reasons Fresnel chose crown glass. First perfected in France, crown glass was hard, light, and easy to mold. It tended to have more streaks than lead glass, but less of an inclination for bubbles. Its low refractive index, though, meant that a prism created from crown glass had to be thicker than one from flint glass. For his lens to work, Fresnel needed blemish-free glass in sizes and shapes that had never been made before.

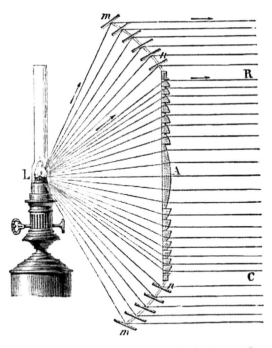

Side view of the Fresnel lens, which bent the light rays from a lamp into a single beam. The sloped sections above and below consist of mirrors, which reflect light that could not be caught in the lens panel.

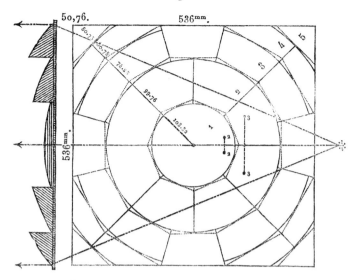

*Drawing that Fresnel submitted to the Commission des phares.
On the left is the side view of the panel, showing the stepped
sections. To the right is the frontal view, showing the arrangement
of the rectangular prisms with which Fresnel was still working.*

To purchase the glass, Fresnel persuaded the Commission to grant him five hundred francs. But his order was not enough to merit the interest of France's state glassworks at Saint-Gobain, which, until recently, had been the *only* place in the country that made glass. Founded in 1665 for the Sun King, Louis XIV, it received a royal privilege and a monopoly on the manufacture of mirrors (which the king loved). The French Revolution ended the monopoly, and a few small competitors now took on modest jobs at low prices. A newly opened company at Choisy-le-Roi, on the outskirts of Paris, was willing to fill Fresnel's order.

The first material from Choisy-le-Roi was disheartening. Fresnel wanted circular lenses that could be placed within one another in a "bull's-eye" pattern. But the glassworks informed him that such custom work would require a larger operation, including grinding powered by a steam engine (a rare technology in 1820). The company could provide him with only the standard polygonal sections. In response, Fresnel modified his design to approximate a circle as best he could with dozens of small polygons. Even these smaller sections were so marred by bubbles and striations, though, that they were virtually worthless. Fresnel feared it might be impossible to make pieces of crown glass that were big enough for his purposes and free of defects.

He turned to François Soleil, the respected instrument maker who had already built several devices for Fresnel and Arago, including for the dramatic light experiment on a circular disc. Fresnel gave Soleil his exacting specifications and subpar materials and hoped for the best. What he received was even better. Soleil figured out a way (still only sketchily understood today) to heat up the glass pieces and remold them so that the defects disappeared. He even learned how to bend the pieces into the arcs that Fresnel wanted, and both men agreed that if they could continue to develop new lenses, their next attempt would involved curved glass.

Of course, they still had to make the first one work. Initially there was trouble with the glue holding the prisms together. Fresnel had first thought to use thick terebenthine of Venice, a standard glue for fastening telescope lenses. But his new lens, with its intense concentration of light, was far too hot. The glue melted and ran out of the joints. Arago suggested some fish glue he had recently received; he had inadvertently discovered its heat-resistant properties when he tried to separate two prisms he had glued by boiling them in water.

Finding the appropriate lamp for his lens presented another problem. The lamp needed to not only emit as bright a light as possible but also be economical and compact. After much study, Fresnel, working with Arago, found that several small wicks gave off a brighter light than a single large wick fueled by the same amount of oil. Multiple wicks, however, made the lamp larger. If it was too large, the lamp would not sit at the focal point of the lens, and any "extra-focal" light would not be bent into the outgoing beam. Fresnel and Arago thus devised an arrangement of four concentric wicks that was small enough, with a nine-centimeter (3½-inch) diameter, for the lens to capture most of the light, resulting in a device that was twenty times brighter than any other lamp then available.

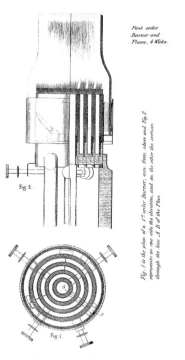

Fresnel and Arago's lamp with four concentric circular wicks.

Left: Fresnel's first flash panel with polygonal prisms, 1820.
Right: Fresnel's first circular glass bull's-eye panel, 1821.

In March of 1820, the "polygonal lens," as Fresnel and Soleil called it, was finished. It consisted of a single panel—a square, twenty-two inches on a side, composed of ninety-seven pieces of glass, each one laboriously remolded and polished, then meticulously placed in a bull's-eye configuration so that each prism bent the light at precisely the correct angle. Its focal length, 27⅗ inches, was remarkably short when compared, for example, with that of a telescope lens, which could be ten feet or more. "Lens power," indicating how strongly a lens bends light, is usually given as the inverse of the focal length (the shorter the length, the stronger the lens). This was one powerful lens, allowing Fresnel to place it very close to the light source, where it would bend the light through very sharp angles.

For his demonstration of the lens, Fresnel gathered together as many members of the Commission des phares as he could. Chairman Sganzin and Admiral Édouard Rossel, a hydrographer who had replaced the deceased Ferregeau, were, Fresnel reported, "dazzled by the spectacle I gave them"—and the single panel, as Fresnel pointed out, was only one of the eight that, in an actual lighthouse, would be arranged in an octagon around the light source. Becquey, so taken with what he saw, asked for seven more identical panels to make up a full apparatus.

A year later, after struggling with insufficient money, defective glass, and the necessity of piecing together polygons into a circle, Fresnel and Soleil had a completed apparatus. Even bigger than the prototype, each of the eight panels was $28\frac{1}{3}$ inches square and able to capture more light. Convinced the light would far outshine anything seen before, Arago arranged for the unveiling of the lens to be a public affair, billed as a competition between Fresnel's design and the recently tested Lenoir and Bordier-Marcet reflectors for the best lighthouse technology available.

The unveiling took place on Friday, April 13, 1821. "It must have been a hardy soul who picked that day," wrote Fresnel with more apprehension than humor. But the evening turned out clear and dark, perfect for observing lights from miles away. Fresnel's lens and its two rivals were set on either side of the Observatoire de Paris. Crowds gathered at the opposite end of Paris, on the hill of Montmartre. Both spots were well matched, elevated enough to afford visibility and so perfectly aligned that the Paris meridian ran straight through them. Indeed, anyone visiting Montmartre today might stumble across one of the bronze discs marking the meridian. Commissioned in 1994 as a monument to replace a statue of Arago that had been melted down in World War II, the medallions are emblazoned with Arago's name and trace the "Arago line," or zero-longitude line passing through the Observatoire, France's rival to the prime meridian, which runs through Greenwich, England.

The audience included the entire Commission des phares, as well as a large number of sailors, invited as the true arbiters of a lighthouse's worth. Lenoir and Bordier-Marcet, the makers of the mirror apparatuses, were also present. What they witnessed, along with the rest of the crowd on the hilltop, was nothing less than the obsolescence of their trade. The Fresnel lens so clearly outshone its competitors that the reflectors now seemed irrelevant. Everyone marveled at the success of the lens. Becquey proclaimed himself "enchanted" with its effect; he saw in the achievement not only the saved lives of countless sailors but also the rare chance for France to beat England in technological advancement. He wanted to send someone straightaway to *Le moniteur*, France's official newspaper, to record the event as soon as possible.

Bordier-Marcet grumbled that the comparisons were "partial and little representative of real conditions"; he sent a personal complaint to Becquey and a fiery letter to the editor of the *Journal du commerce* (which published a glowing review of the lens), but he could not deny that Fresnel's light was brighter. If the lens overtook the parabolic reflector, by qualities that he admitted he did "not hesitate to recognize," then the entire technology of reflectors would "enter into oblivion."

Fresnel was not surprised by the success of his lens. It merely demonstrated the inevitable outcome of his calculations. But he admitted in a letter to his brother, "I don't know if it's my state of malaise which renders me almost insensible to the success of my lens; I would think rather that it's that I did not have the pleasure of surprise, having realized in advance, by the measurement of shadows, the effect that it would produce." Nevertheless, Fresnel, no longer a provincial engineer with a few visionary ideas about optics, became France's great hope to prove its technological superiority. The next lens he built would not be the result of patchwork experiments funded largely out of his own pocket, but an advanced device intended for France's crown jewel of a lighthouse, Cordouan.

"Versailles of the Sea"

Cordouan was the oldest continually operating lighthouse in France. It stood alone on a small outcropping of rocks at the mouth of the Gironde estuary, which fed the two major rivers, the Garonne and Dordogne, running through France's rich wine country. Bordeaux, then France's third largest city and its richest port, depended on this waterway for its trade, but more than a few ships had crashed on the rocks, which were visible only at low tide. Michel de Montaigne, the Renaissance essayist and mayor of Bordeaux, persuaded King Henri III to secure funds for a lighthouse and was present in 1584, when workers started laying stones. Cordouan was one of the great architectural gems of France. Louis de Foix, the most celebrated architect of the sixteenth century, planned it. He spared no detail, giving the four-story structure a richness of decor more typically reserved for palaces. The base was a massive entrance hall, adorned by fountains and busts of Henri III and his successor Henri IV. A grand staircase led to the second story, designated "The King's Chamber." Louis XIV had the room refurbished in case he stopped by. He never did, but the opulence of his quarters, with its two fireplaces, sculptures, fine marble, and medallions, made the phrase "Versailles of the sea" more than just a nickname. The third story was the chapel, complete with an altar and four exquisite stained-glass windows. The top floor was an empty room intended for the fire. It was, incidentally, the only room the keepers regularly visited, for they hauled the fuel up a separate "servant's entrance" staircase alongside the main tower.

When the light was finally lit in 1611, during the time of Louis XIII, it was pretty simple: flames from a large pile

The lighthouse at Cordouan, seen from the sea.

An interior view of the tower at Cordouan.

of burning wood. But as France's most notable lighthouse, Cordouan was often the first to experiment with new innovations, initially with pitch and tar as the fuel, then coal in the 1720s. In 1782, the lighthouse used France's first oil lamps and mirrored reflectors, although the latter were rather primitive, not even parabolic in form. Sailors angrily reported that the light was virtually invisible and demanded a return to the open coal fires. The Ministère de la marine sent an engineer, Joseph Teulère, to investigate. He replaced the reflectors with larger versions and taught the keepers how to clean them without damaging their reflectivity. Still the complaints rolled in from the sea, and in 1788, Teulère decided to start from scratch. He had the top level of the lighthouse torn down and rebuilt taller, to raise the light and give it a longer range. The new lantern room that housed the apparatus had enormous (fifteen-foot-high), opulent glazed windows. The reflectors were the finest France could offer: Lenoir's twelve 30-inch parabolics, rotating on a triangular frame so light flashed every minute. Yet the visibility of the light remained disappointing, and sailors still pleaded for the old fires.

Fresnel's lens represented another opportunity to make Cordouan a showpiece for French technological capability. Fresnel and Soleil were determined to improve the lens. In particular, Fresnel wanted to switch from the rectangular panels they had been using to approximate the desired shape, to curved, or annular, pieces of glass. From an optical standpoint, the advantage of a curved lens was huge. It largely corrected the lingering problem of spherical aberration—the tendency of a lens to bend rays more strongly at its edge than at its center—which had been exacerbated by the use of rectangular prisms. With annular pieces, this aberration could be virtually eliminated by reducing the number of segments needed and, ideally, making the prism in one big ring—an extraordinary demand for glassmakers.

Again Fresnel called on Arago, who argued the point to Becquey, who asked the company at Choisy-le-Roi for eight annular prisms, each cast either as one piece or in four to six pieces. But each piece the glassmakers made exploded as they worked it. Finally, Jean-Pierre Darcet, director of the company, gave up and suggested that Fresnel try the glass manufacturer Saint-Gobain. Although the great glassworks had lost its official royal privilege, it nonetheless retained a de facto monopoly on high-quality and specialty pieces. Saint-Gobain had previously shown little interest in Fresnel's requests, but it could scarcely refuse the head of the Ponts et chaussées himself. Indeed, the director of Saint-Gobain, Benjamin-Marie Tassaert, took particular care of the order, overseeing the process as the glass was poured into specially made molds.

The prisms now arrived in great arcs, quite close to Fresnel's specifications. The vigorous final polishing, done by groups of women, made the edges of the lenses so sharp that Fresnel and Soleil had to blunt them to avoid injury during assembly. But with the larger prisms—nineteen inches long and two to three inches thick—even the best glassmakers could not keep them free of defects. Fresnel carefully inspected every one he received, and work proceeded slowly as a good portion of every batch of prisms had to be returned.

In addition to Soleil, Fresnel worked with a small team of assistants. The only official one was Jacques Tabouret, another young engineer at the Ponts et chaussées. Fresnel was happy for the help, but his two most trusted associates were in fact unpaid and unofficial: his brothers. Léonor had followed in what now seemed like a family tradition: attending first the École polytechnique, then joining the Ponts et chaussées.

Under Becquey, he began to specialize in canal construction and spent a good bit of time near the river Oise not far from Paris. In May of 1825, Becquey placed Léonor at the disposition of the company in charge of building a canal from Paris to the sea, which allowed him to be near his brother until 1827, when he was officially appointed Augustin's assistant. Fulgence, in contrast, broke with family tradition. A good seven years younger, he showed no interest in engineering whatsoever. His great passion was languages. He came to Paris to study Chinese, but he helped Augustin where he could, usually with contracts and negotiations, like the diplomat he would eventually become.

The apparatus they were building for Cordouan was the biggest and most complicated yet. In addition to the bull's-eye panels he had been using for his demonstrations, Fresnel added eight small additional trapezoidal lenses, which fit above the others like a sort of roof, with a hole in the top serving as a chimney. Any light that would have been lost through the top was thus bent straight upward, where it ran into eight large, plain silvered mirrors that fanned out around the center. At the bottom, he arranged 128 small mirrors to catch any light that escaped from below, layering them "like the slats on a Venetian blind." These mirrors reflected beneath the big lenses, thereby adding a fixed light to the turning light at no additional cost. The lens had a focal length of thirty-six inches.

It took a year to obtain enough glass meeting the specifications and then to assemble it into a working lens. By July of 1822, Fresnel had moved the apparatus into the Arc de Triomphe and was ready to test it. Becquey postponed the demonstration until Louis XVIII returned from a month's absence. Fresnel passed the time with a little unofficial testing of his own: occasionally, he admitted to his brother, "a good number of honest Parisians" out for a late-night stroll on the Champs-Élysées suddenly found it lit up as if it were daytime.

For the official test, on August 20, 1822, the Commission des phares joined the royal entourage at a cathedral in the bucolic village of Mortefontaine, twenty miles north of Paris. The almost new moon was a stroke of luck, leaving a deep, dark night for the viewers. Fresnel fretted that the filmy haze

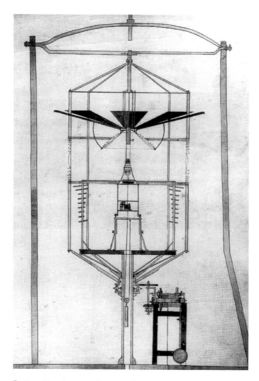

Drawing of the Cordouan lens. The main beam went out from the central bull's-eye panels, drawn in profile here. The bottom "Venetian blind" arrangement and the fan on top were mirrors that reflected the light that could not be captured in the bull's-eye panels. An escapement mechanism, to the right, rotated the lens.

Photograph of the Cordouan lens, which is currently on display at the Musée des Phares et Balises in Ouessant, France.

left over from the day's heat might disperse the light, but the beam, now rotating, was as blinding as ever. Fresnel noted that he was also able to see the constant fixed light produced by his arrangement of mirrors, though it was weak compared to the brightness of the rotating beam. There was never a moment when the light was completely dark, however.

The king and the Commission alike pronounced the demonstration a success. With testing over, Fresnel packed up his apparatus and sent it to Bordeaux, where it was kept until the appropriate time for installation. A ship could land on the tower's rock only under the calmest of weather conditions, and the area was notorious for its terrible winter storms. The keepers at Courdouan could go months without contact with the shore, and were usually given six months' worth of food and supplies at the beginning of winter in case they did not make it back to land until spring. The new light would have to wait for winter to pass.

During this time, Fresnel poured his energy into a period of jaw-dropping scientific productivity. Between September of 1822 and January of 1823, he wrote a series of papers that gave the theory of light its most substantial overhaul since Isaac Newton, developing into a comprehensive theory his original insight that light behaved like a wave. There was a certain frantic element to his work, an effort to commit everything down to paper quickly. His health had always been precarious, but in the winter of 1822/23 it began to deteriorate severely. Rather than diminishing his output, however, this seems to have spurred him on, desperate to see the foundations of his legacy established. Each paper virtually laid the framework of an entire field, as he gave complete accounts of double refraction, circular polarization, and the propagation of light through crystals. Fresnel's vision of light and its physical propagation would serve as a framework for much of the physics of the nineteenth century. Thomas Young, looking to advance the wave theory in Britain, recruited Fresnel to write the entry for double refraction in the *Encyclopaedia Britannica*, to counter the corpuscularian bent of David Brewster's optics entry. Fresnel, citing his health, deferred the task to Arago.

With spring, Fresnel's health improved, and he headed to Cordouan to oversee the installation of the lens. The lighthouse had hired local men to assemble the light, but they were, quite understandably, not well acquainted with the demands of precision optics. "Having almost only bad workers with me," he wrote to Young, "I was obligated to enter into the minutest details of this operation, and frequently to play the workman myself." Tabouret came down from Paris to help him with the optics, while Bernard-Henri Wagner, who supplied the rotating machinery, sent an assistant, Monsieur Hans, to help with the mechanical part, an escapement mechanism typical of the clockwork at the time. They spent the month of July on the tiny rock outcropping, an episode that Fresnel did not describe fondly in his letters. By the end of the month, though, they had their light in place. On July 25, 1823, the world's first Fresnel apparatus, at Cordouan, was in use.

Fresnel visited the local ports to gauge its reception. Royan, the town on the mainland directly north from the lighthouse, was full of sailors, while Verdon, to the south, housed a number of military officers. Both groups marveled at the brilliance and whiteness of the light. Some English bathers at the seaside also affirmed that it was much better than anything they had seen in England. Fresnel worried that the praise was exaggerated and wrote to a friend in The Hague to try to get a sense of what Dutch sailors thought of it.

But Fresnel had little reason to doubt the excellence of his light. By his own calculations, the apparatus at Cordouan produced about as much light as thirty-eight of the best English reflectors, yet consumed only half as much oil. As Fresnel wrote to Robert Stevenson, head of Scotland's Northern Lighthouse Board, "It's only by following the experimental method that I have just exposed to you, that is, by *measuring* the intensity of the light in all directions that one can compare with exactitude the relative merit of the different lighting systems." It was almost impossible to make a comparison based on how far off the light could be seen because, unlike that from other lighthouses, the beam remained bright right up to the point where it disappeared from the horizon. Given the curvature of the earth, Cordouan's light eclipsed at nine nautical miles for sailors on low-lying piloting sloops (even though, at sixty-four meters, or two hundred feet, the lighthouse tower was tall). Sailors up in the rigging, however, could see the light, still entirely bright, at a remarkable thirty-three nautical miles away.

The French government had spent twenty-eight thousand francs on the Cordouan lens, or roughly seventy thousand dollars in today's money. But unlike the parabolic reflectors it replaced, this time the investment paid off. Sailors noticed immediately, and approved of the change. Word spread quickly across the seas.

Fresnel had every reason to congratulate himself, and his scientific reputation soared. The year 1823 saw his election to the Académie des sciences. The next year, the government bestowed on him the Ordre national de la Légion d'honneur (National Order of the Legion of Honour), the highest decoration in France, invented by Napoléon to recognize exceptional duty to the country. But Fresnel continued to work, plotting improvements to the lens arrangement, the lamp, the turning mechanism. He had made the dioptric lens based on refraction a reality, but his work was just beginning.

THREE

THE DREAM OF TOTAL REFRACTION

EVEN AS THE CORDOUAN LIGHT flashed into operation,
Fresnel was dreaming of a better device. The painstakingly molded prisms of his bull's-eye panels had done
what he had wanted: precisely bending the rays into a
central beam. But they were able to capture only about half of
the light since much was lost through the top and bottom of
the lamp. The laws of optics did not allow the prisms to bend
light through anything more than a forty-five-degree angle, so
the light that radiated from the top and bottom simply could
not be bent to join the central beam. This necessitated placing
mirrors above and below the prismatic panels to reflect the
light that could not be refracted.

Fresnel soon conceived of a more elegant solution by
exploiting another optical trick. Light traveling through a
prism with the correct angles would hit the second (internal)
surface of the prism at a sufficiently sharp angle that it would
reflect back inward before finally leaving through the third
side of the prism. Such an "internally reflecting" prism would
thus act like a mirror and redirect light through a wider range
of angles than a simple refractive prism allowed.

Unlike mirrors, which lost half of their light at the point
of reflection, this prism would lose nothing at all. Fresnel

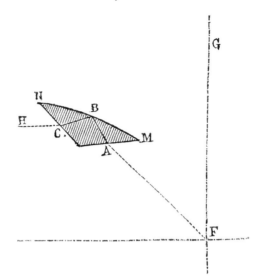

*An internally reflecting prism. A ray of light entering at point A
would hit the internal surface of the prism at point B,
and be completely reflected, exiting the prism at point C.*

envisioned an arrangement that placed these more angular
prisms above and below the conventional bull's-eye panels
to form what he called a catadioptric apparatus, combining
both reflection (the catoptric part) and refraction (the dioptric part). He successfully worked through the calculations,
complicated by the different geometries of the refracting
and reflecting prisms, but once again discovered that no glass
manufacturer could construct what he wanted: a reflecting
prism of significant size. He put together a small version,
a few inches high, which was used to guide ships along the
canal Saint-Martin (an extension of the canal de l'Ourcq,
his former project, into Paris). It proved the advantages of
the arrangement, but the perfect lighthouse lens still hung
tantalizingly out of reach.

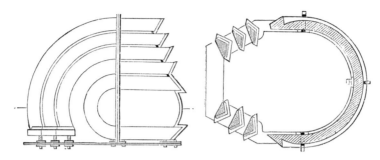

Fresnel's first catadioptric lens, placed along the quays
of the canal Saint-Martin.

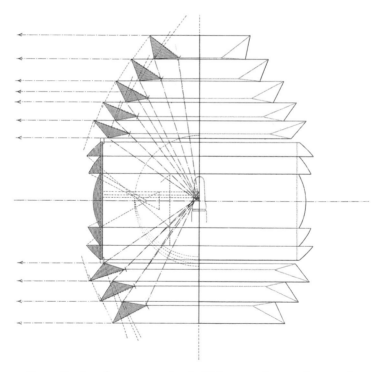

Fresnel's plan for a two-and-a-half-foot catadioptric lens, with
internally reflecting prisms arranged along the top and bottom.

THE CARTE DES PHARES

Meanwhile, however, France embarked on a massive project to mark the entire coastline using Fresnel lenses. Becquey promoted Fresnel from his job as temporary assistant, to secretary of the Commission des phares, the first member of the Commission to carry a salary. Together with Admiral Rossel, and in the consultation with sailors, Fresnel began to develop a plan. Since they knew how far away the light from the dioptric lens could be seen, they were able to calculate

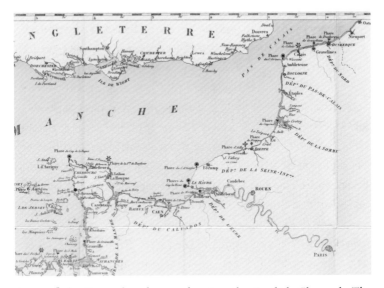

Inset of the Carte des phares, showing the English Channel. The orders of the lenses are represented by the size of the circles (the large ones are first-order lenses; the small ones, third order; there is only one second-order lens shown here, at Carteret). Fixed lenses are repre-sented by solid circles, while rotating lenses have marks indicating the number of panels on the outside.

where lighthouses should be placed along the coast so that as one light faded out of sight, the next came into view. The grand plan that Fresnel revealed in 1825 named the Carte des phares, or Lighthouse Map, called for a system of fifty-one lighthouses—only thirteen were currently operating in France—and all would have new Fresnel lenses. Conceived as a rational network in which everything fit together, the plan divided the lights into categories based on size. The largest were "land lights," the first beacons sailors would see coming in from the open sea; these would be roughly eighty miles apart. Between them would be smaller lights whose specific characteristics would help sailors identify where they were and help guide ships into particular harbors.

The classification system Fresnel devised divided the lenses into first-, second-, and third-order lights, and a fourth category of smaller harbor lights. First-order lights were the largest, with an interior diameter of nearly 2 meters (6½ feet). Second-order lights measured 140 centimeters (4½ feet) across; third-order lights, 50 centimeters (1½ feet); and the harbor lights, 30 centimeters (1 foot). The different-size lights had different lamps as well, with the largest using Fresnel and Arago's innovative four-wick design; the second-order light using a three-wick lamp, and the third-order a two-wick lamp. The larger lights thus consumed a great deal more oil. Most of the lights in the plan—twenty-eight—were of the first order. The five second-order lights were for particular areas that did not need the brightest signals, while the eighteen third-order lights were reserved for use in sounds, river entries, and bays.

Sailors had to be able to distinguish the lights from one another to know where they were. Fresnel proposed three variations: fixed lights, which shone continuously; rotating lights that flashed every thirty seconds; and rotating lights that flashed every minute. The light flashing every minute used eight bull's-eye panels, rotated once every eight minutes, and shone the brightest. The thirty-second flashing light had sixteen half-panels, which divided the light into twice as many (weaker) beams. Fixed lights, the weakest of the three, had no bull's-eye panels but sent their beams evenly across the horizon.

Mapping out the lights, placing the brightest ones where they were most needed while ensuring that no two adjacent ones were alike, was a challenge, similar to piecing together a jigsaw puzzle.

While Fresnel worked on the Cartes des phares, France's second Fresnel lens was installed—at Dunkirk, the northernmost port. A severe winter storm had wiped out the city's only lighthouse in the winter of 1823. With Dunkirk as the center of cod operations in France, sending off some 130 boats carrying thousands of sailors every March for the six-month fishing season, the city leaders cast around for a way to keep the cod doggers off the treacherous sandbars surrounding the harbor. The highest point in town was the tower of Leughenaer, but as part of the original fortifications of 1548, it had been erected to defend against cannons rather than to radiate light. The Ministère de la marine advised against using the structure to house the light, and warned of the danger it posed to large ships, which would already be in the sandbars by the time sailors saw its beam. This advice, however, was issued just before the Fresnel lens at Cordouan demonstrated how much brighter it made the light. Becquey felt free to ignore the ministry's warning, and a third-order fixed lens constructed by François Soleil for seven thousand francs was installed by February 1, 1825.

In Dunkirk for the installation of the lens, Fresnel also attended a celebration inaugurating new flood gates recently installed in the port. There he ran into a clockmaker, Augustin Henry, who was mounting a new system of carillon bells for the town. Fresnel was looking for an experienced clockmaker to replace Wagner, who had provided the machinery that rotated the lens at Cordouan. Wagner's device controlled the rotation by ticking through the teeth of a gear wheel, a system that Fresnel found unacceptably jerky and uneven. Fresnel found his man in Dunkirk.

Henry came from a long line of eminent clockmakers. His father, Pierre Henry, was one; his mother's family name, Lepaute, had been synonymous with quality clockmaking for generations. Pierre Henry died when Augustin was six, and afterward his mother placed him with a relative who trained him in the craft. (Augustin eventually took the Lepaute name himself when he married Anaïs Lepaute, a distant cousin, in 1834.) The young Augustin impressed Fresnel with his energy and ideas and soon began supplying the mechanical components for Fresnel's lenses, replacing the jerky escapements with a centrifugal fan regulator, which rotated smoothly and continuously.

With the Dunkirk light up and running, Fresnel set out to visit each site designated on the Carte des phares. The Commission des phares hoped to delegate much of the work of building the dozens of new lighthouses to the members of the Corps des ponts et chaussées stationed in the coastal departments. In each location, Fresnel met with the *arrondissement*'s (district's) ordinary engineer, who would in theory be in charge of lighthouse construction. But the novelty of the technology meant that Fresnel had to stay actively involved. For example, he had compiled some preliminary notes for the lighthouse of Goulphar at Belle-Île and sent them to Monsieur Luczot, the chief engineer of the region. But Luczot ignored the notes, claiming that Fresnel's walls were too thin. Fresnel insisted, drawing up his own design and submitting it to the general board of the Ponts et chaussées, which unsurprisingly sided with him. Luczot still refused to use Fresnel's design, even when Fresnel, extremely annoyed, went back and explained in detail the calculations supporting it. The board more or less forced Luczot to follow Fresnel's plan.

Even when lighthouse towers already existed, the installation of Fresnel lenses was not always easy. Barfleur, on the Normandy coast, was supposed to get the next lens out of Soleil's workshop, the one after Dunkirk's. The Barfleur lighthouse, dating from 1774, guarded the site of one of Europe's most famous shipwrecks: the *White Ship*, which sank in 1120, drowning the only legitimate heir of King Henry I of England and setting off the crisis known as "The Anarchy." But after Soleil completed the first-order lens for Barfleur, the engineers discovered that the existing lighthouse was not wide enough to hold it. So the Barfleur light went to Île de Planier, a new tower completed in the spring of 1826, and the Île Planier lens, the next to come out of Soleil's workshop, was sent to Belle-Île. Construction began on a new tower for Barfleur; a representative of King Charles X presided over the laying of the first stone.

Soleil was swamped. He had several different orders going at once, and his workshop was a sprawling profusion of prisms waiting in stacks for assembly. Sometimes the towers were completed but the lens was not ready yet. When the coastal engineers for the Granville Lighthouse, a high-priority site in Normandy, asked for its lens, Fresnel promised to finish grinding the curved glass himself because Soleil was too busy. The Commission des phares made it clear in an 1826 report that it wanted to have more than one lens manufacturer involved, both to speed up production and to provide some healthy competition. But the Commission had a hard time finding anyone besides Soleil who was willing to take on the task. Only one man, reputable instrument maker François-Antoine Jecker, even attempted, submitting a small harbor lens as a sample. Fresnel, however, found his prisms inexact, with a radius of curvature of 4.527 meters when Fresnel had asked for 4.4 meters. Fresnel demanded high standards. He had complained in 1825, "It is quite difficult to give the taste for exactitude to those who do not have it, especially when they are older and lack an elementary knowledge of geometry!"

Fresnel now had a "control workshop" of his own, where he could run experiments and try out new techniques. With the field of photometry in its infancy, Fresnel struggled to quantify the brightness of the light coming through his lenses, eventually settling on a comparison to an equivalent number of Carcel lamps (the standard lamps of the day) burning oil at a set rate of forty-two grams per hour. By 1825, he noted, the lenses had the brightness of more than three thousand Carcel lamps.

By 1827, eight employees worked in Fresnel's workshop, under the direction of two Ponts et chaussées engineers, Fresnel's longtime assistant Jacques Tabouret and a Monsieur Boulard. Their chief preoccupation was to develop the catadioptric lens. The prisms needed to fit together in extraordinary precision for the lens to work. The slightest change of degree would send light rays out at the wrong angle, never to join the rest of the beam.

Glass production remained a problem. Without sufficiently high-quality glass, the lenses would not work well. Fresnel had managed to get some decent material from Saint-Gobain when the minister of interior had made his projects a national priority. But Fresnel needed to outfit every lighthouse in France, and Saint-Gobain's output was slow. Fresnel again tried Choisy-le-Roi, but he returned shipment after shipment as unusable. His letters showed his mounting frustration: "I waited some days to answer you, wanting to first assure myself that the material of your last casting was as full of bubbles and streaks as it appeared to me initially. For that, I cut and polished two facets on three pieces selected among those where the glass appeared the least bad, and I recognized that this matter was not resolved. I would expose myself to the just reproaches of my Administration in accepting glass so full of defects." Saint-Gobain also repeatedly delivered prisms with too many defects, so Fresnel asked Benjamin Marie Tassaert, the director of Saint-Gobain, always to send a few extra pieces to make up for ones that had to be discarded. Even when the glass was free of bubbles, it was not always cast to the right degree of precision. "Monsieur Soleil complains, with reason, that your workers are not careful enough in the casting of his rings. I noticed at his place a large number of glass pieces that cannot serve, although they contain good enough material, because they were poorly cast," Fresnel wrote.

The problems highlighted the virtually unmeetable demands of the dioptric lens. On the one hand, the scale and comprehensiveness of the Commission's plans required mass production of the new technology. On the other hand, each lens required precision craft, dependent on traditions in glassmaking and optical instrumentation better suited to specialty markets.

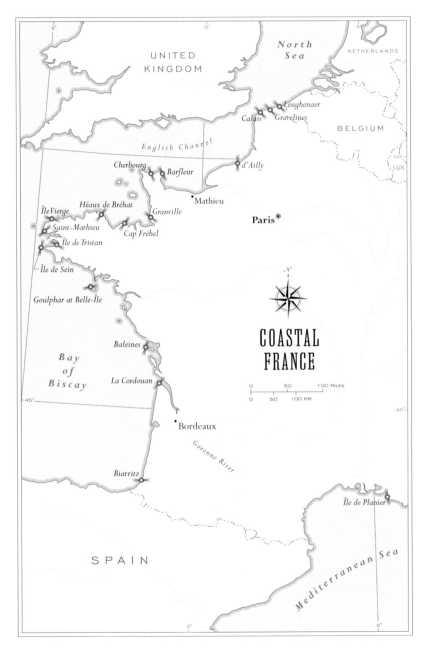

The coast of France with Courdouan, Barfleur, La Heve, Dunkirk,
Calais, Goulphar, Ailly, Granville, and Baleines.

In the 1820s, the Industrial Revolution was scarcely evident in France. Steam power, its engine and hallmark, was rare. Still, Fresnel was struck by the idea of using steam to help speed up production. All of the lathes in Soleil's workshop (and there were many of them) were turned by a single horse. On December 17, 1826, Fresnel wrote to Jean-Louis Roard, a friend of his uncle Mérimée's and a chemical manufacturer at Clichy, to ask if he could use the factory's steam engine: he proposed placing a grinding machine, only a square meter (ten square feet) in area, on the floor of the chemical factory and powering it by steam. Roard was interested, but the Ponts et chaussées wanted a several-year commitment from Fresnel before it set up the new work space, and Fresnel, admitting he had "neither the leisure nor the health," pulled out.

Fresnel could not ignore the transformations being wrought by steam engines across the channel. Britain's Industrial Revolution, with several decades of momentum, was changing the country into a landscape of factories and steam power. Fresnel watched nervously, well aware of the implications. The faster, more powerful steam-turned lathes could produce bigger, stronger annular prisms. It was his own inability to make such pieces that left his vision of the catadioptric lens unfulfilled.

It was thus with some apprehension that he watched his friend and colleague François Arago leave for England in 1822 with a lens in his bags. Arago planned be there for a few weeks to continue measurements of the meridian north to England, working with his brother-in-law Mathieu to coordinate the French and British meridian measurements. The British had measured the meridian as far north as the Orkneys, and the French as far south as Ibiza. They now proposed to link these projects across the English Channel, using the same technique of triangulation that Arago had used in 1806 with Jean-Baptiste Biot: shining light across distant points and measuring the angles. In place of the parabolic reflectors he had used in Spain, he carefully packed one of the large, three-foot dioptric test panels. Fresnel wrote to Mathieu begging him to ensure that the famously effusive Arago remain discreet on the subject of the lenses. "If one questions Arago on the process that we use for the construction of annular lenses, it would be good, I think, if he does not give too many details on the subject. If the English get mixed up in it, they will perhaps quickly pull ahead."

It is easy to understand why Fresnel was concerned. In addition to England's undergoing an astonishing period of industrialization, its storied seafaring tradition had already given the country more lighthouses than any other. The English lighthouse authority dated back to 1514, when Henry VIII founded a guild of "the most glorious and Undividable Trinity of St. Clement," more commonly known as Trinity House. All of the lights were built under private contract: Samuel Pepys, master of Trinity House from 1676 to 1689, and best known today for his diary, acknowledged "the evil of having lights raised for the profit of private men, not for the good of the public seamen, their widows and orphans." Nonetheless, private lighthouses continued to flourish, each exacting astronomical "private dues" from ships in a process that resembled extortion. By 1800, the situation was scandalous, and ship owners demanded a more regular and just system. Trinity House, accordingly, began buying up the private lighthouses in a process that was slow, contentious, and costly (more than a million pounds). Expensive new equipment was therefore not on Trinity House's agenda when Arago journeyed to England. In truth, Fresnel had nothing to worry about.

Yet Fresnel's concerns were not entirely misplaced. The British engineer assigned to work with Arago, Colonel Thomas Colby, could not help but notice the new technology. After setting up the lens at Fairlight on the English coast, they crossed the channel to Cap Blanc Nez, France, where, even though forty-eight miles away, it shone "like a star of the first magnitude." Colby, duly impressed, mentioned what he saw to his good friend Robert Stevenson, who happened to be in charge of all the lighthouses of Scotland at that time.

Despite a jagged coastline that teemed with hidden dangers, Scotland had nothing resembling a lighthouse authority until late in the eighteenth century. Before that, the entire nation had been a coastal no-man's-land. Sailors avoided it if they could; one early act made it a crime for ships stocked with essential goods to leave Scottish ports between October and February. One lone light, a coal fire at the Isle of May run by Scottish courtiers since 1635, scarcely halted the mounting casualties.

As merchant trade increased, the number of shipwrecks rose accordingly. The situation came to a head in 1782, when a particularly bad set of winter storms took out much of the Scottish naval and merchant fleets. Ship owners and captains pressured the government to do something. Parliament finally acted, and in 1786 it set up the Northern Lighthouse Board to oversee the not-yet-existing lighthouses of Scotland.

Thomas Smith, a former lampmaker, was the Board's first engineer. He relied heavily on his sturdy, capable stepson, Robert Stevenson, who by the age of nineteen was building his own lighthouse on the River Clyde. In 1797, Stevenson succeeded Smith as the engineer to the Lighthouse Board. Unlike his stepfather, who apparently disliked spending months at a time touring barren rock outcroppings, Stevenson relished the numerous journeys around the coast. He was soon famous throughout Britain, both for his ingenious designs and for his brushes with danger. Completion of the masterful Bell Rock tower in 1810, the tallest offshore lighthouse of the time, which he designed and built, cemented his reputation.

Soon, Stevenson was on his way to France to see Fresnel's lenses for himself. By the time he arrived, in August of 1824, Cordouan was in operation. Stevenson decided to spend the first two weeks in Paris, then head to the coast to see the lens in action. On Sunday he strolled through the Père Lachaise cemetery, which had recently become a popular tourist site after gaining the remains of the star-crossed lovers Abélard and Héloïse, and on Monday attended a demonstration of the lenses at the Académie des sciences. In the following days he met Fresnel himself, who, at Becquey's request, happily showed Stevenson around. Their conversation was somewhat hampered by the fact that they did not speak each other's language. As they parted, Stevenson presented Fresnel with his book on the construction of Bell Rock, with a handwritten dedication to "Monsieur Frenel [*sic*], membre de l'Institut, with the author's compliments." The gift perhaps invokes the tenor of their meeting: friendly, full of goodwill, and lacking a certain understanding.

Stevenson next toured the coast, with Cordouan the chief object of his interest. He admired the brightness of the light and took careful measurements of the lens's dimensions. On his return to Scotland, he wrote to Fresnel and ordered two large annular lenses and a four-wick lamp. After compiling a report on his findings, Stevenson presented them to the Northern Lighthouse Board on December 31, 1824. He recommended use of the new technology, with caution: he wanted to see how the lenses would work in practice. Stevenson set up an arrangement at Inchkeith, then removed it to make modifications. He performed many experiments over the next several winters, when weather prevented him from touring the coast, but he and the Board continued to delay the installation of one of the lenses in a lighthouse.

One source of delay was the involvement of an increasingly deranged David Brewster. Stevenson knew Brewster from Edinburgh social circles, and no doubt thought the expert in optics would be interested in this bit of technological novelty. Little did he know that the name "Fresnel" was already quite familiar to the man, and not in a good way. Brewster had continued to rail against the wave theory of light long after everyone else had accepted it, his often vicious attacks against the rival "undulationists" leaving him increasingly marginalized and excluded from university positions. Some fault can surely be put on his difficult temperament ("fractious to the extent of something like insanity" was how one colleague put it), but Brewster himself blamed Fresnel and his adherents. He was particularly embittered about the Royal Society of London, Britain's premier scientific organization, which would have nothing to do with him. The entire body, he claimed, was corrupted by Fresnel's theory, which functioned as "the Creed of the Society" and poisoned its members against him.

Needless to say, Brewster's reaction was not the delighted appreciation Stevenson had hoped for. The very object of his ongoing paranoid fixation had trumped him once again. Brewster's immediate response was to claim he had come up with the idea first, pointing to an encyclopedia article he had written in 1811 suggesting the use of "polyzonal lenses" as burning glasses for concentrating heat. Arago wrote a blistering reply, refuting Brewster's claims and decrying the "bizarre disposition of spirit" that made him incessantly claim other people's ideas as his own.

Brewster next designed his own dioptric lens, a more complicated affair using roughly the same principles as Fresnel. He tried to interest Robert Stevenson in it, but when Stevenson kept ordering lenses from France, Brewster turned to the Royal Society of Edinburgh in 1827, condemning the incompetence of Fresnel, Stevenson, and the entire French system. He demanded that the Northern Lighthouse Board appoint him to it and place him in charge of introducing dioptric lenses. When the Board declined, Brewster contracted the London instrument makers W. & T. Gilbert to make a lens of his own design, bankrupting the company in the process—it went under in 1828.

Brewster's challenge to Fresnel never got very far, and his design never made it into a lighthouse. But Fresnel had been right to worry that once the British learned of his work, they would try to outdo him. His dream of a perfectly reflecting apparatus remained elusive, hampered by the physical limits of glass production. And by 1827, it was beginning to be too late.

FOUR

RACE TO PERFECTION

J UST AS FRESNEL'S LIGHTS began flickering on along the French coast, his health took a turn for the worse. His cough, already nearly permanent, had started to produce mucus tinged with blood shortly after Cordouan was lighted. There could be little doubt that he had consumption—tuberculosis—the leading cause of death in nineteenth-century France. The disease was ubiquitous but poorly understood. Doctors generally dismissed the notion that it was contagious, and attributed it instead to an individual's constitutional predisposition, in conjunction with an attack of "*les passions tristes*," the sorrowful passions. Few diseases lent themselves so well to the period's Romantic view of suffering as tied to an exquisite sensitivity. The consumptive heroines of *La bohème*, *La dame aux camélias*, and *Les misérables*, all of whom grew more saintly and ethereal as their death approached, were not simply literary fantasies, but reflections of the widespread understanding of how the disease worked. Fresnel fit the reigning medical storyline of a sensitive and creative soul whose mental forces outpaced his physical ones. His life, too, seemed headed to apotheosis, as everything but the lighthouses fell away. He gave up his research on light and his teaching responsibilities for what he considered his pressing duty to France.

As the weather turned colder in 1826, the cough worsened and even lighthouse work became too much for him. His brother Léonor began to assume more of the responsibilities (Fulgence had left Paris for Rome the year before, to take up the study of Arabic). The winter stretched out, with a thick layer of snow and mud covering the streets through February. In March, Fresnel wrote to Becquey asking to hire Léonor as his assistant. For the past two years, Léonor had been working on a canal connecting Paris to the sea, although he helped his older brother every chance he got. With the canal work over, Léonor was concerned he might be sent farther afield. Becquey quickly agreed to the request, acknowledging the "deplorable state" of Fresnel's health.

Fresnel hoped to return home to Normandy in the spring, but his vastly deteriorated health made the journey too risky. He was carried instead to a country house in Ville-d'Avray, a small town between Paris and Versailles bucolic enough that the early impressionist Camille Corot would later make his reputation painting its charming fields and riverbanks. Fresnel's mother traveled down to nurse him. Yet the fresh spring air had little impact. That month, the British Royal Society awarded Fresnel the prestigious Rumford Medal for the most important contribution to science of the last several years (signaling Britain's conversion to Fresnel's new physics). Arago had come to deliver the medal, only to be shaken on finding his friend so deathly ill. Fresnel, Arago reported, scarcely looked at the medal. Instead, he thanked Arago in a soft whisper for what he supposed was a painful task, noting that "the most beautiful crown means little, when it is laid on the grave of a friend."

Fresnel, thirty-nine, died in his mother's arms eight days later, on Bastille Day—July 14, 1827—the anniversary of a revolution he did not approve of but which had nonetheless shaped his life. He was remarkably philosophical about death, which he had seen coming for a while. His only regret derived from his ever-present sense of duty. Science had so much to offer on questions of public utility, "I could have wished to live longer, perhaps I might have had the happiness of finding the solution to some [of these questions]."

LÉONOR TAKES OVER

Within a few days after Augustin's death, Léonor received two letters from Becquey. The first expressed the director's regret; the second offered Léonor the now vacant position of secretary of the Commission des phares. Léonor's response combined gratitude for Becquey's "paternal kindness" with an insistence on honoring the "noble heritage" of his brother by continuing it. Promising to "force himself to overcome the pain to satisfy the duties of the position," Léonor stepped into the shoes of his older brother. He was to hold the position for twenty years and oversee the completion of the plan Augustin had sketched out.

Léonor shared with Augustin his delicate features, high forehead, sharp cheekbones, and tapered, gently clefted chin, as well as an identical education and a similar ambition as a civil engineer. But Augustin's near reclusive shyness stood in contrast to his brother's easy friendliness. Augustin, from his first days on the road crew, had hated the duties of administration. Léonor found his own form of genius within it. The pair had long ago divided responsibilities, with Léonor shouldering the managerial work his brother found so onerous. Now, with the comprehensive Carte des phares under way, managerial skills were exactly what the Commission des phares needed.

The system Léonor took over in 1827 was a bit of a mess. Most of the lighthouses, and their new, improved lenses, were still only plans on paper. Along the coasts of France, the new and refurbished towers stood in varying states of completion, while back in Paris, Soleil's workshop was spilling over with lenses in various stages of assembly.

The design of the lenses continued to evolve. Léonor inherited, along with Augustin's other positions, the direction of the control workshop that tested new ideas. Its focus remained the catadioptric lens, Augustin's beehive clustering of reflecting and refracting prisms, which the lensmakers could not get to work on a large scale. Their efforts were hindered by the limits of the available technology and the frustrating lack of access to a steam engine.

For the Fresnel lens to reach its potential, it needed not just a single brilliant insight, but the transformation of an industry. When Soleil and Fresnel began working together, their partnership echoed a long tradition of experimental collaboration in place since the seventeenth century: the physicist and the instrument maker devising ingenious methods of teasing out nature's secrets. Optical instrument making had long been one of the great craft traditions, requiring the utmost skill and finesse. Soleil was a master at it, and the lenses he produced were glittering testaments to his patience, labor, and skill. But France's massive plans for the rapid expansion of lighthouses called for something else, a revolution in the process of industrialization. As the Industrial Revolution started to rumble through France, Soleil, for one, was ready. Immediately after the success of Cordouan, he opened the world's first factory for lighthouse lenses. The contrast with his old optical instrument workshop, which continued to operate, was striking. At 35 rue de l'Odéon, the old shop nestled in the heart of Paris's academic scene, among the city's major universities and a short walk through the Jardin du Luxembourg to the Observatoire. The new enterprise rose in the rapidly industrializing suburb of Saint-Denis. This small town to the north of Paris had been home to the first Gothic cathedral and the burial site of virtually every French king since the seventh century. By the nineteenth century, however, the town had become best known for the number of factories moving in. The canal Saint-Denis opened in 1824, connecting the town to Paris and the Seine and providing easy transportation of goods. Soleil moved in the next year, on the forefront of the industrial boom.

The glass industry was undergoing its own transformation. In many ways, it was emblematic of the French style of industrialization, characterized by much stronger government involvement than what was seen with the English model. The state, from Napoléon onward, viewed its well-trained engineering corps as its best bet for catching up with England. The glass industry exemplified the new mixture of state and free-market principles in business: Saint-Gobain went from being a royal enterprise to a private corporation, becoming fully divested from the state by 1830, but it was run by civil engineers trained at the École polytechnique, the national engineering school. Intent on modernizing, glassmakers improved both the quantity and the quality of their lighthouse prisms.

Industrialization received a further boost in 1830, when the decidedly conservative Bourbon regime succumbed to revolution once again, this time replaced by the king's younger cousin, a pudgy fellow whose regime was best known for the admonition *"Enrichissez-vous!"*—Go get yourself rich! Nicknamed the Bourgeois King, he was a friend to business and industry. Augustin Fresnel had a posthumous hand in the revolution, in a sense, courtesy of François

Arago. Well on his way to replacing Laplace as the dominating personality of French science, Arago had been elected permanent secretary of the Académie des sciences (the body's most important position) on July 7, 1830. One of his first official duties was to give Fresnel's eulogy. The morning he was scheduled to deliver it, however, King Charles X announced that he was revoking the Charter of 1814, which acknowledged the monarch's contractual obligations to the people of France (effectively, he declared absolute rule). Arago, skirting censorship laws, used the eulogy to subtly attack the king's actions, emphasizing Fresnel's commitment to the Charter and warning for the future of the regime. The speech, widely circulated, was proclaimed "a triumph" by Alexandre Dumas, and forty-eight hours later "three glorious days" of fighting erupted that put Louis-Philippe on the throne.

A hero of the democratic left, Arago soon combined his scientific career with a political one. His election to the Chamber of Deputies in 1830 added a great friend of lighthouses to France's legislative body. "Modern lighthouses," he claimed on the Chamber floor soon after his election, "are one of the most beautiful conquests of science." He pointed to the lives saved by the Fresnel lens: the average number of shipwrecks had dropped from 163 a year to 107 after its introduction.

The lens needed a friend in the Chamber. The Carte des phares project had run out of money and was languishing. Arago reminded his fellow deputies of both the economic toll of shipwrecks and the unconscionable fate of the nearly five hundred French sailors, "our most useful, our most courageous citizens," who died in them every year. The Chamber voted an extra two and a half million francs. The Commission des phares eventually added Corsica, Algeria, and other

French colonies to its purview; Arago rhapsodized about the lens's new reach. "One day," he told the Chamber, "the globe will present the spectacle of an immense illumination, which will facilitate even more the circulation on the ocean, this great route of all nations."

THE GREAT ROUTE OF ALL NATIONS

Arago championed the potential of the lens, envisioning the increased circulation of people and trade as a boon to all nations. And the lens soon went global, as other countries ordered their own. Norway was the first to do so, installing one on the island of Oksøy in 1832. The Netherlands was next, ordering a second-order light for the island town of Goedereede, and placing it, in 1834, at the top of the local church's tall, square steeple, which had guided ships since 1552. Scotland, despite its early interest, was still debating what to use, having only recently set up a major comparison between a French dioptric lens and British reflectors (the audience, gathered twelve miles away at the astronomical observatory on Calton Hill, pronounced in favor of the lens). Ten years had passed since Stevenson's visit to Fresnel, and the Northern Lighthouse Board wanted another fact-finding mission to France. This time the lucky traveler was Robert Stevenson's oldest son, Alan, groomed virtually from birth for the lighthouse business.

The thoughtful, introspective young man had leaned toward the study of literature at the University of Edinburgh, and worried, in a letter to his father, that his talents were not those of an engineer. Robert, horrified to discover that the twenty-one-year-old had been writing poetry in his spare time, warned him to devote himself to studying technical works with no distractions. And study Alan did. Of a decid-edly delicate constitution, he was no match for his father when it came to tromping around the remote seacoasts. But he was a gifted student, strong in math and science, with a supple imagination, precisely the talents demanded by mod-ern engineering.

Alan, twenty-seven years old, arrived in Paris in the spring of 1834. Léonor, whom he had already met during a tour of Europe, welcomed him warmly, and the two developed what was, from both accounts, a genuine friendship. Léonor intro-duced Alan to Soleil and showed him around the workshop, pointing out all the new developments, including the use of internally reflecting prisms. Alan spent nearly every day for over a month either in Léonor's office or at the workshop with the assistant Tabouret, writing home about the latest improvements and their "notable advantages." He contin-ued, "All, however, that I shall really have to do is to give an account of what has been effected by the late illustrious Fresnel, who seems to have devoted such minute attention to every detail of the Dioptric apparatus, that he has foreseen and provided for almost every case that occurs in the practice of Lighthouse illumination."

Alan, who understood the implications of Augustin Fres-nel's work more clearly than his father did, returned to Scot-land intent on introducing lenses there. The first one went up almost immediately, replacing the reflectors at Inchkeith in 1835. The next year, he installed another lens at the Isle of May, all the while planning for even more significant changes. He remained the whole time in close contact with Léonor, and invited him to Scotland to supervise the progress.

Eager to travel to Britain, home of the Industrial Revolu-tion, Léonor petitioned the Ponts et chaussées to finance a trip, claiming that it was essential to see the applications of steam power. He would be bringing his new bride, Eulalie Françoise Réal-Lacuée, whom he had married in 1836. (Becquey, the director general of the Ponts et chaussées, and Léonor's cousin and close friend, Prosper Mérimée, who was fast becoming one of France's most celebrated young writers, were witnesses.) They set out on their British tour less than a year later. She had lived in exile in America during her childhood, and her fluent English supplemented his more workman-like language skills. The new couple had a grand time in Scotland, touring not only the coast but also the picturesque Scottish Highlands. Madame Fresnel, known for her voice, made a particular impression on the Stevensons by singing parts of an opera before breakfast one morning.

Amid the revelry, though, was a serious education. Léonor met the lighthouse authorities of Scotland and England and inspected their new lens installations. The most enduring impression, however, came at Cookson glassmakers, which had, in consultation with Léonor, made the lenses for the lighthouses at Inchkeith and the Isle of May. Lacking the trade secrets to duplicate France's high-quality glass, Cookson imported its glass for the lenses and struggled to make a profit (it ultimately went bankrupt in 1845). Léonor, however, was impressed with one thing: the factory had been completely outfitted with steam-powered lathes for grinding the glass. Their use resulted in an important breakthrough, as Cookson had just completed the first form for the central portion of a fixed dioptric lens (called the drum) consisting of a single, circular piece of glass, improving the quality of the lens considerably. This was long thought to be impossible, and Soleil, with his horse-turned lathes, still made them in several pieces.

Achieving the Dream

On his return from Britain, Léonor began pushing French manufacturers to modernize. He encouraged his clockwork supplier, Augustin Henry, now using the name Lepaute, to try his hand at constructing the lenses, awarding him the commission for a second-order lens for the Santander Lighthouse in Spain. The Commission des phares had long wanted more than one manufacturer to produce the lenses, and Léonor did all that he could to support Lepaute, assigning a few Ponts et chaussées engineers to help with the technical details, arranging use of the new glassworks at Prémontré, and encouraging Lepaute to experiment with a steam-powered lathe.

In 1838, the same year Lepaute completed the Santander commission, François Soleil retired, turning over the family's long-standing optical instrument business to his son, Jean-Baptiste François Soleil (usually referred to as François Soleil Jr.) and the lighthouse business to his daughter Adelaide's husband, Jean Jacques François. François and Lepaute were now competitors, although their relationship was quite good and they often collaborated. François, too, began using steam-powered lathes. Léonor presented the situation as "a happy concurrence," where the two craftsmen spurred one another on to greater heights by a process of "noble emulation." He did everything he could to keep a balance between the two companies, distributing the orders as evenly as possible between them. There was more than enough work to go around, and Léonor hoped a little competition might spur them on to create the first large-scale catadioptric apparatus.

The interior of Soleil's factory. The bands visible along the sides are used to transfer the power of the central steam engine. The image is from the 1850s, after the company was sold to Louis Sautter.

The exterior of the lenswork factory.

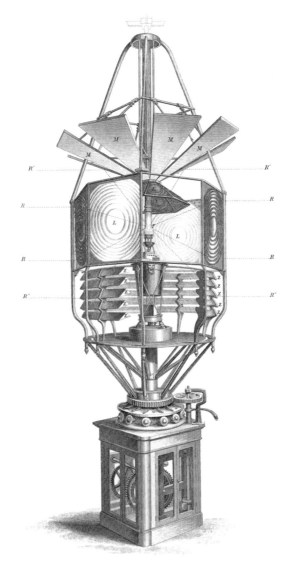

The Skerryvore lens: a "crown of glittering glass" in the words of
Robert Louis Stevenson's "Skerryvore: The Parallel." The lens retained
Cordouan's upper fan of mirrors, but the lower slats were replaced with
a ring of internally reflecting prisms.

Once the two companies began using steam engines to turn the lathe, they could start making the curved annular pieces larger. This was the crucial step for making a catadioptric lens. The central portion of the fixed dioptric lenses had been large, with diameters of nearly two meters (6½ feet). But these could not be produced in a single piece, and were sometimes composed of as many as thirty-two smaller, curved sections fitting together to form the ring. In theory, a catadioptric lens could be made using the same principle, with the rings composed of several different pieces. In practice, it did not work though, as the pieces had to be too finely adjusted. The successful catadioptric lenses, such as those on the canal Saint-Martin, had internally reflecting prisms made from a single mold, and they were small, no more than twenty-five centimeters (sixty inches) in diameter—the limit of what hand-turning could do. With steam power, however, François produced catadioptric rings one meter (3¼ feet) in diameter, which in 1842 he placed in two third-order fixed apparatuses, for the lighthouses at Gravelines and Île Vierge. These were the first major catadioptric lenses ever used, but they were still not the largest, first-order ones. Léonor declined to order any of these, claiming their execution remained "grave and perilous."

Alan Stevenson was bolder. Having been fascinated by the catadioptric dream since his time in France, he hoped to harness the internally reflecting prism for his own monumental masterwork currently under construction at Skerryvore. His eighth lighthouse and by far his greatest engineering challenge, its elegant tower was the tallest in all of Britain and sat on half-submerged rocks completely exposed to the unbroken force of the Atlantic Ocean. He resolved to crown it with the best-possible lens that science could provide. He needed a first-order revolving one, however, and no existing machinery could produce the top portion's enormous annular catadioptric rings in a single piece. But Alan was, as François later recalled, "determined to put in practice a plan which I had long contemplated, of placing totally reflecting zones below the lenses." While retaining a Cordouan-style fan of mirrors around the top, the design for Skerryvore replaced the bottom arrangement of mirrors with five annular rings, their triangular shape carefully designed to catch the light radiating downward, reflect it off the bottom surface, and refract it back out to the horizon. Both François and Lepaute vied for the job, submitting large annular catadioptric rings one meter (3¼ feet) in diameter as samples. François wound up with the contract, which called for annular rings of an unprecedented two-meter (6½-foot) diameter. Léonor gave François the necessary technical drawings but warned him to "heavily weight the consequences of the engagement he was going to take." François made each ring in four separate arcs. After providing the molds to Saint-Gobain, which made the rings, François ground and polished them himself on a steam-powered lathe.

Lepaute, meanwhile, garnered his own major contract: England's venerable lighthouse at Eddystone, celebrated in sea shanties around the world. Trinity House had just completed the process of buying up private lighthouses in 1842, and immediately turned to outfitting them with lenses. They sent Michael Faraday, a newly appointed "scientific advisor," to France, where he met with Arago, Lepaute, and Léonor Fresnel on successive days. After touring the factory where the Eddystone lens was under construction, Faraday recorded in his diary, "Went to Mr. H. Laponte [sic], at his workshops. He now both makes the optical and the mechanical part, and his workshops and apparatus are excellent, far surpassing anything we have in England." Although smaller than the one at Skerryvore, being only a second-order light, the Eddystone lens went a step further: the mirrored headdress was eliminated and a "cupola" of totally reflecting prisms was added on top. Léonor Fresnel spent a good bit of time in the Lepaute workshop, where he "experimented with the greatest care" with the prisms, all arranged in large rings, with their angles carefully computed. Once installed, the lens shone brilliantly, a hundred times brighter than the fire of the twenty-four candles that first lit the Eddystone tower in 1759.

Both François and Lepaute exhibited their innovations at the Exposition de l'industrie (French Industrial Exposition) in 1844, where they won gold medals. François displayed a lens there so exact that it collected light on a surface seventy centimeters (28 inches) wide, and concentrated it into a beam of four millimeters (.15 inches). The two men's successes accumulated, arriving also in lockstep precision. When Arago submitted a note to the Académie des sciences praising François, he made sure there soon was another one highlighting the skills of Lepaute. When one won the Légion d'honneur in 1844, the other soon found himself decorated as well. That same year, François passed the Soleil company down to his son-in-law, Theodore Létourneau, who changed the name to Létourneau & Cie. The firm's close relationship with Lepaute remained, however. The two collaborated, occasionally producing lighthouse equipment together. And they competed, too, most pointedly to be the first to make a completely catadioptric first-order lens.

This last, final step was no easy matter. Léonor Fresnel retired from the Commission des phares in 1846 to compile his brother's collected works. He was replaced by Léonce

The construction site at Héaux de Bréhat. Each piece of granite for the tower was taken from a quarry on a neighboring island, assembled there to test it, then disassembled, with each piece carefully numbered like a giant puzzle. About sixty workers crowded the tiny island and lived under miserable conditions. They slept in a dormitory of ten by three meters (thirty by nine feet), with two levels of hammocks and no toilets. Workers were allowed only one bath a month ("obligatoire," it specified, in case that seemed too frequent for anyone). There was soon an insurrection, and Léonce Reynaud had to pay prime wages to convince the stonecutters to finish. The whole project cost three times its estimate, although it did produce a very fine lighthouse, bearing the "sublime simplicity of some gigantic ocean tree." Its first-order lens was lit in 1840.

Reynaud, another Ponts et chaussées engineer, who had recently overseen construction of the most ambitious and arduous lighthouse the French had ever built, the one at Héaux de Bréhat. But despite knowing adversity, Reynaud still hesitated over the difficulties of making a first-order catadioptric. He first required assurance that the lensmakers themselves would cover the considerable investment in new equipment. He then ran a series of tests in 1850 to ensure that a completed apparatus, which would be an enormous eight feet tall, would not be too heavy for the wheel system on which it would rotate. He finally placed the orders in 1851, one with Lepaute for the lighthouse at Ailly and one with Létourneau for the lighthouse at Saint-Clément des Baleines. These were two of the most important and conspicuous lighthouses in France and, as it happened, practically the last lighthouses in the country without Fresnel lenses. Their towers were among the most solidly built of the pre-1800 generation (Ailly had been built in 1775, Baleines in 1682), but neither of them could easily accommodate the new system. What was more, the Commission des phares had just updated them with brand new reflectors almost immediately before the technology became obsolete (in 1822 for Ailly, in 1820 for Baleines).

Léonor had in fact ordered a lens for Ailly in 1840, and it sat while the engineers first tried to modify the old tower and then decided to build an entirely new one. By 1843, however, the fan of mirrors around the top of the lens already looked out-of-date. Not wanting such an important lighthouse to contain outmoded material, Léonor sent the lens elsewhere. In its place, Lepaute designed a new first-order lens like no one had ever seen before. The top part consisted not of simple annular rings, but of eight separate panels, each of which focused the light into a beam. The light from these beams was set about four degrees ahead of the light from the main dioptric bull's-eye panels. An observer would thus first see the light from the crown panels, with a brightness of fifteen hundred Carcels (as the standard of comparison was now known), and then, as that light began to dim, the light

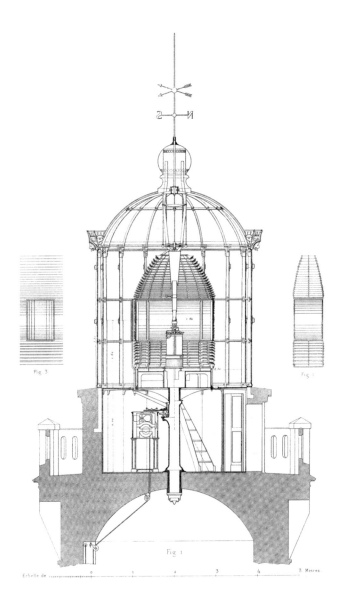

*A fixed catadioptric lens, shown installed
in a lantern room. The arrangement of the rings
of prisms gave it the nickname "the beehive."*

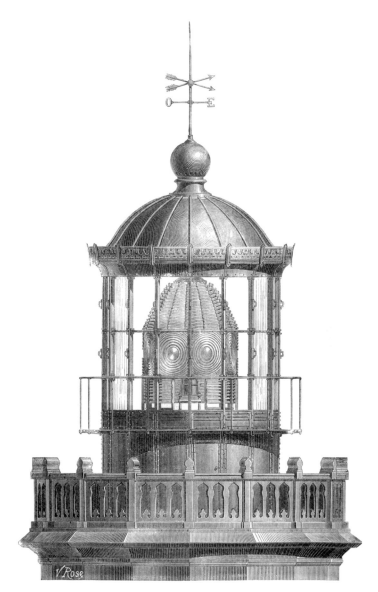

A rotating catadioptric lens, with eight dioptric panels.
The catadioptric section on top bends light into the main beam,
while the bottom section sends out a constant fixed light.

from the main bull's-eye, four thousand Carcels bright. This increased the duration of the flash from seven seconds to fourteen seconds (the primary complaint about the new lenses did not concern their brightness, but the briefness of the flash). The apparatus also retained its lower annular rings, the light from which remained continuously visible, even though not as bright. The arrangement was considerably more complicated than anything tried before. Reynaud sent an engineer to install it in September of 1852, warning that "the installation of a 1st order apparatus is a long and difficult operation."

Létourneau, too, threw himself into the construction of his own first-order catadioptric for the Baleines Lighthouse, a painstaking enterprise involving the casting of enormous curved pieces and precise coordination of roughly a thousand prisms. He finished it in 1853, losing the race to Lepaute but winning the title of the first fully catadioptric revolving lens. To achieve this, he had replaced the bottom segment of fixed annular rings with revolving sections that directed the light toward the main beam.

The lighting of the Baleines lighthouse on January 15, 1854, marked the culmination of Augustin Fresnel's vision in at least two ways. It was the fulfillment of his long-delayed catadioptric dream, which had taken more than twenty years and an industrial revolution to achieve. It was also the final lighthouse completing Augustin Fresnel and Admiral Edouard de Rossel's 1825 rational plan. Fresnel lenses now illuminated the entire French coastline, with their studied differences of flashing and duration allowing sailors to pinpoint their location exactly. France's premier historian, Jules Michelet, praised the tight network of overlapping rays for its enlightened design: "For the sailor who steers by the stars, it was as if another heaven had descended to earth."

THE HOLOPHOTAL CHALLENGE

France sought to highlight its achievement—the perfection of the catadioptric lens—at the first Exposition Universelle it hosted the next year, although this would also become the occasion for the most strenuous challenge to French lens supremacy from Britain yet. The Exposition had been conceived as France's response to London's Great Exhibition of 1851, which had showcased the impressive fruits of British industry and offered a particular slight to French lens manufacturers. The 1851 Exhibition's most eye-catching piece of lighthouse equipment, a first-order lens with a revolving drum and fixed annular zones above and below, displayed in the British section, bore the name of the British glassmakers Chance Brothers. It was designed and constructed by Jacques

The Crystal Palace at the Great Exhibition of 1851 prominently displayed the Chance Brothers lens designed and built by Jacques Tabouret.

The original plans for the Exposition Universelle of 1855 called for a Fresnel lens to occupy the central position (shown here being set up) in the Palais de l'Industrie, although it was later moved to the side.

Tabouret, Fresnel's former assistant, who had been recruited to the company by Georges Bontemps, himself a political refugee from France. The jury's praise was modest, pronouncing the construction admirable but the glass inferior to French glass. And Chance Brothers had yet to outfit any actual lighthouses. Regardless, the British rivalry was keenly felt.

In 1855, the French were determined to reclaim their source of national pride. A first-order revolving fully catadioptric lens—the object of perfection everyone had been striving for—took center stage at the Exposition, with the original plan placing it in the middle of the main hall of the Palais de l'industrie. Although later moved to the western side to accommodate the massive crowds (over twenty-two thousand people visited every day for the first 198 days the exhibit was open), the lens still drew the visitor's eye, mounted on a tall column graced with a bust of Augustin Fresnel. Behind was an enormous fresco by the painter Jean-Léon Gérôme, rendering homage to Fresnel. Allegorical depictions of sixteen different countries, from Norway to Brazil, flanked either side of the inventor, all bearing the dates on which they adopted their first Fresnel lens.

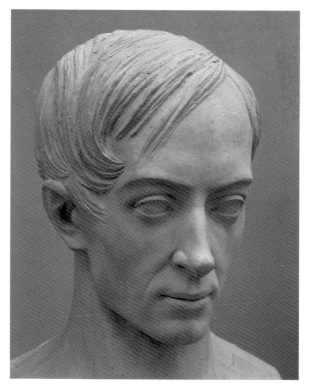

Bust of Fresnel, displayed at the 1855 Exposition.

By this time, French manufacturers had delivered more than two hundred lenses to various foreign powers around the world. The guide to the Exposition stressed the lens's role as France's gift to humanity. Its manufacture was an "eminently national industry," which showed France in its best light:

> The invention of these devices, due to a French engineer, encouraged and developed by the public administration, brings to a very high degree the imprint of the particular nature of our spirit and general tendencies, for it was deduced from considerations of a purely scientific order, conceived outside of any private speculation, in view of general interests, and classed immediately in the number of benefits for humanity. It is one of our least contested titles to the recognition of civilized peoples.

Two of the features that separated French industrialization from its English counterpart were its strong contingent of scientifically trained state engineers and its lesser dependence on private investment, both crucial factors in Fresnel's achievement. The lens, deduced from scientific principles and alleviating a near intractable source of human misery, perfectly exemplified the values of enlightenment, civilization, technology, and industry that France wished to highlight in the Paris exposition.

Yet just at the moment of crowning victory, another challenge came from Britain. In a letter to the Exposition's jury, Thomas Stevenson, the youngest of Robert's sons, put forward the claim that he deserved credit for the lens's most perfect form. Thomas had overseen the completion of the tower at Skerryvore after Alan left in 1842 to replace his father as the chief engineer for the Northern Lighthouse

Board. It was thus Thomas who fitted the lighthouse with François's lens, and in the years that followed he became convinced he could do better. Too much light, he argued, was being wasted in the lower section, which sent light out in all directions. To counter this, he designed what he called the "holophotal system" (from the Greek for "whole light"), which directed all of the light into the central beams.

Thomas Stevenson billed the idea as the ultimate perfection in lens design. Not everyone was happy with the claim. Brewster, with characteristic petulance, responded that he should be given credit for being the first one to use the term "holophote." The French, with perhaps more reason, replied that they had been working toward precisely that goal for years. What they saw as a Scottish grab for credit entailed a fair bit of confusion and sometimes deliberate obfuscation. Thomas had first introduced the idea in an 1849 talk, "Account of the Holophotal System of Illuminating Lighthouses," but had apparently applied the term "catadioptric" to devices that were essentially old-fashioned parabolic mirrors with lenses in front, rather than internally reflecting prisms. While not technically wrong ("catadioptric" simply means using both reflection and refraction), it was certainly misleading. As was Thomas's claim that an apparatus he built along these principles for the Horsburgh Lighthouse in 1850 made it "the first lighthouse in which total reflection was applied to revolving apparatus." Given that the lighthouse was in Singapore, and he never provided any drawings for it, he seemed to be counting on the fact that no one would notice that the reflection in question was provided by nine parabolic mirrors, rather than prisms.

Thomas did add a few paragraphs explicitly referring to internally reflecting prisms when he ultimately published his paper in 1851. But by this time, Reynaud had already ordered from Lepaute the panels for the Ailly Lighthouse, which were based on a similar principle. It was true that the Ailly lens was not technically "holophotal," since the prisms directed the top light not into the major beam, but slightly ahead of it (to increase the length of the flash rather than its brightness). But it was also clear that the French were familiar with internally reflecting prisms. The real fight concerned the events of 1854, when the French lit a fully catadioptric light in Baleines on January 15, and the Scots lit one in North

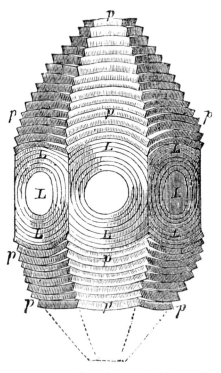

Thomas Stevenson's design for a holophotal lens, which would bend all of the available light into the central beams.

Ronaldsay on September 18. Thomas claimed, however, that he had designed the lens for North Ronaldsay in 1850, commissioned it from Létourneau in 1851, and received it in 1852, making it the truly first catadioptric revolving light (the delay until its lighting in 1854 was due to problems with the tower).

French authorities were quick to respond. "The glory of the invention of lenticular lighthouses, due to the illustrious engineer Fresnel, was contested from France by the Scottish engineer, M. Stevenson," reported the *Annales de ponts et chaussées*. Léonor Fresnel wrote an open letter to Arago, published by the Académie des sciences, defending his brother's early advocacy of internally reflecting prisms. Reynaud went even further, claiming that Thomas Stevenson must be lying, because when Reynaud first asked Létourneau to construct a catadioptric section for the Baleines lens in 1852, the lensmaker had neither the mold nor the tools to make it, implying he had not just finished one for the Scots. Reynaud wrote a long open letter to the jury of the Paris exposition, arguing that French engineers had already conceived most of Thomas's improvements. "As for the catadioptric mirror and the adjective holophotal," he concluded, "I give them to him with no contest; and I will even formally swear never to use either one of them." Despite borrowing a bit of Stevenson's design in his later work, he never did.

There was little basis to Stevenson's claims for perfecting the lens. Augustin Fresnel had seen the same theoretical potential of internally reflecting prisms early on, but had been unable to translate the vision into full-scale apparatus. It was, if anything, the lens manufacturers, glassmakers, and myriad architects of the Industrial Revolution who ultimately delivered the perfect lens, each large curved piece a marriage between optical theory and the hard reality of glass.

The rebuff from the French was not the biggest disappointment of Thomas Stevenson's career, a moment that came when his only son told him he wanted to be a writer instead of a lighthouse engineer. "You have rendered my whole life a failure," Robert Louis remembered his father yelling at him. Thomas had dragged the boy to worksites on remote northern skerries in an effort to get him interested in lighthouses, but all that would come of it were vivid settings for the shipwreck scenes of his later novels. Robert Louis, who became one of the most celebrated poets and novelists of his time, fled not just the family business, but Scotland as well, first to France and then to California, Hawaii, Tahiti, and Samoa. His travels, while wide, kept him close to the sea, and he admitted later in life that "whenever I smell salt water, I know that I am not far from one of the works of my ancestors."

The family business lingered in his poetry as well, including an 1887 ode to the Skerryvore lighthouse:

> . . . But there
> Eternal granite hewn from the living isle
> And dwelled with brute iron, rears a tower
> That from its wet foundation to its crown
> Of glittering glass, stands, in the sweep of winds,
> Immovable, immortal, eminent.

The crown of glittering glass had only grown more ornate after Skerryvore was built, as its top row of mirrors gave way to ever-more complex arrangements of prisms. More than one thousand prisms could go into a single catadioptric revolving lens, each carefully tailored to its role in directing the light into a single beam. By 1851, Lepaute and Létourneau switched entirely over to the new, perfected design. That same year, Augustin Henry-Lepaute retired, placing his company in the hands of his two sons (who retained the company name). The next year, Theodore Létourneau, the third generation to run the Soleil company, sold the firm to an outsider, the engineer Louis Sautter. The two companies maintained much the same relation, however, as Léonce Reynaud tried to distribute the work between them equally. And by then, there was more work than ever before, as countries all over the world began ordering lenses of their own.

FIVE

THE AMERICAN EXCEPTION

B Y THE MIDDLE of the nineteenth century, the Fresnel lens had transformed not only the coasts of Europe, but those of Africa, Asia, and the Americas as well. Cuba, Brazil, the Bahamas, even tiny Tobago had lighthouses with dioptric lenses. The great exception was their neighbor to the north. The United States, despite some tentative early experimentation, remained stubbornly resistant to the new technology. Foreign observers often noted with puzzlement the American reluctance to adopt the Fresnel lens. The country was, after all, astonishingly dynamic. While Europe was sorting itself out after the Napoleonic wars, the American merchant fleet rose to become the most prominent in the world. Between 1790 and 1810, the tonnage American ships carried tripled. American ports were hives of activity, sending off the products of a flourishing economy undergoing a "market revolution."

America's vitality was undeniable, but its ingenuity was long in supply and short in scientific understanding. With pure science often couched as elitism, American engineering had a decidedly utilitarian bent. Alexis de Tocqueville, who toured the country in the 1830s, wrote that "hardly anyone in the United States devotes himself to the essentially theoretical and abstract side of human knowledge. . . . Everyone is on the move, some in quest of power, others of gain. In the midst of this universal tumult, this incessant conflict of jarring interests, this endless chase for wealth, where is one to find the calm for the profound researches of the intellect?" The very bustle of American life, for Tocqueville, was incompatible with the studied contemplation of pure science.

In France and Britain, the Fresnel lens had ridden in on the wave of a new technological ethos. The lenses had succeeded because of a highly trained engineering force, well grounded in the principles of science and mathematics. A professionalized, state corps had replaced apprenticeships and family traditions. A generation of lighthouse builders, hands-on men whose training took place more often on rocky coasts than in classrooms, had begrudgingly made way for university-trained engineers.

The education that Augustin Fresnel received at the École polytechnique, or what Alan Stevenson had at the University of Edinburgh, was simply not available in America in the early nineteenth century. Colleges, still struggling to be taken seriously, were notoriously poor in teaching science and mathematics. No schools offered formal instruction in engineering. The first one to do so, the United States Military Academy at West Point, copied the curriculum of the École polytechnique course for course. Sylvanus Thayer, the superintendent who introduced the program in 1819, had spent nearly two years in France studying engineering; he brought back professors, lesson plans, and a central textbook written by none other than the Commission des phares's Joseph-Mathieu Sganzin. A new generation of West Point engineers would work hard to gain acceptance of Fresnel's lens, but an entrenched old guard remained deeply skeptical of the theoretical benefits of fancy French devices.

Chief among the skeptics was Stephen Pleasonton, the head of the United States Lighthouse Establishment. An accountant with his eye fixed on the bottom line, he simply could not understand why anyone would pay five thousand dollars for a lens when he could buy a full reflecting apparatus for less than a thousand. He expressed no interest in how the lenses worked or how they compared to reflectors. His incuriosity deepened into outright hostility as an avalanche of complaints about American lighthouses accumulated from sailors, merchants, and insurers—everyone interested in keeping ships afloat. Pleasonton continued to insist, against all available evidence, that the U.S. lighthouse system was the best in world.

LEDGER BOOKS AND OIL LAMPS

The United States had some dozen lighthouses when it declared its independence in 1776. These were, for the most part, local operations, paid for by charging light-dues to the ships using the harbors. The first federal Congress took control of them, however, in the Lighthouse Act of August 7, 1789 (setting a precedent in invoking the Commerce Clause). Responsibility for lighthouses bounced back and forth between the secretary of the Treasury and the commissioner of the revenue, with neither particularly interested in the job. Indeed, through the administrations of Washington, Adams, and Jefferson, the president himself made virtually all major decisions regarding lighthouses. Only in 1820, during James Monroe's presidency, was a specific office, the United States Lighthouse Establishment, created when the position of the commissioner of the revenue, who happened to have jurisdiction at that time, was reorganized out of existence. Although conferring a grand title on the office, the change reflected just bureaucratic shuffling. It gave responsibility for lighting the country's coast to the fifth auditor of the United States, a position to which Pleasonton had just been appointed.

Pleasonton, born in the same year as his nation, was a sober and serious young man. The sole surviving photograph of him shows his oversized mouth pulled down into a disapproving grimace, and his long neck ramrod straight. He had neither the countenance nor the temperament of a dashing war hero, yet he owed his position to precisely such a reputation. In the War of 1812, he had single-handedly saved the Declaration of Independence. Far from being a soldier, he spent the war as a minor clerk in the State Department in Washington, D.C. As the British began advancing on the city in 1814, President James Madison tasked him with moving the department's records. Pleasonton bought bolts of linen, made them into bags, threw into them as many documents as he could— including the Declaration of Independence, the journals of Congress, George Washington's correspondence, and various papers of the State Department—and loaded them into a cart. General John Armstrong, the secretary of war, observed the scene from across the hallway in the War Department and dismissed the frantic activity as the result of "unnecessary alarm," adding that he did not believe the British would actually make it to Washington. Pleasonton persisted, however. The plan was to deposit the records at an abandoned gristmill on the Potomac River, but Pleasonton drove his cargo to a safe location in Leesburg, Virginia, some thirty-five miles away.

Pleasonton described what happened next: "Being fatigued with the ride, and securing the papers, I retired early to bed, and was informed next morning by the people of the hotel where I stayed, that they had seen, the preceding night, being the 24th of August, a large fire in the direction of Washington, which proved to be a light from the public buildings the enemy had set on fire, and burned them to the ground." The British had burned the White House, the Capitol, the Treasury, and nearly every other government building. The State Department was nothing but smoking rubble, and the only records that survived were the ones on Pleasonton's cart.

The young clerk received a hero's thanks for his efforts. Yet it was a distinctive heroism: marked by prudence, meticulousness, and going to bed early. It shows Pleasonton at his best,

Stephen Pleasonton, fifth auditor of the United States.

full of loyalty and conscientiousness. Yet when he brought the same fastidious caution, and the same insistence on holding onto relics from the past, to his responsibilities overseeing lighthouses, the country would not be well served.

The official description of Pleasonton's new position as fifth auditor lacked any mention of lighthouses. Rather, as the State Department's official accountant, his job was to review the department's books and make sure they were in order. The fifth auditor was thus an auditor. (Yes, there were auditors first through fourth, responsible for checking the books of customs, the army, the War Department, and the navy, respectively.) For Pleasonton, the lighthouse obligations entailed another set of books to review. Here is how the system worked: Pleasonton delegated most duties to the customs collectors who had lighthouses in their districts; they were named superintendents of lights and were responsible for building and maintaining the lighthouses. Pleasonton reviewed and approved their expenditures. (Before 1822, they had to submit all expenditures; after 1822, they submitted only those exceeding a hundred dollars.) And here the accountant in him shone: he was ruthless in keeping expenses in line.

Pleasonton had never been to sea, or even lived near it. He had no scientific training or engineering knowledge. If he seems a poor choice as the overseer of the coast's great technological marvels, bear in mind that at the time there was nothing technological or marvelous about a lighthouse: it was a tower with a fire on top. The earliest American lighthouses were lit by candles, and when the federal government took over, the lamps comprised nothing more complicated than a pan of oil with at least one wick stuck into it. (The most advanced one, with up to eight wicks, was nicknamed the "spider lamp.")

The lamps were simple, but the oil was expensive. At first, the lamps were lit with whale oil, made by rendering blubber, which had the inconvenient property of congealing in the cold. Lighthouses had separate supplies of "summer oil" and "winter oil," the latter being thinner and more expensive (and still liable to congeal if it got cold enough). By Pleasonton's time, the government had switched to non-congealing sperm oil, culled from somewhat mysterious organs inside the sperm whale's head. It gave the clearest, brightest flame available, but also cost four times as much as the more widely used whale oil. Almost the entire expense of maintaining a lighthouse was the cost of supplying it with sperm oil. The ledger book, with the price of sperm oil per gallon carefully weighed against the rate of consumption, was the Lighthouse Establishment's chief object of interest, and the State Department's meticulous accountant seemed well suited to its inspection.

When Pleasonton took charge in 1820, the United States was enjoying a glut of cheap sperm oil, brought on by the discovery, in 1818, of the immensely lucrative "offshore grounds" for hunting whales deep in the Pacific Ocean. From that point on, ships ranged widely across the ocean, from the South Seas to the Arctic. It was not long, though, before overfishing depleted the stocks, and by the 1830s, whale oil prices were rising again, soon to the point of crisis. In fact, Pleasonton's tenure coincided with one of America's most dramatic boom-and-bust cycles, as the price of sperm oil increased fourfold.

American lighthouses were closely tied to the fate of the sperm whale. France did not use the product—its lighthouses relied exclusively on colza oil, squeezed from the seeds of the turnip-like rapeseed plant, which grew plentifully in France, its fields of brilliant yellow flowers common throughout the countryside. Britain had also switched to colza oil as the price of sperm oil began to climb. Only the United States, which controlled the market for whale products, stuck firmly by sperm oil.

The ledger books spoke their implacable logic: the only way to keep the costs of lighthouse maintenance down was to burn less oil. One man promised to do just that. Ex–sea captain Winslow Lewis, creator of his own patented lamps, guaranteed to illuminate lighthouses using half the oil of any of his competitors, and to charge less while doing it. For the cost-cutting Pleasonton, it was a match made in heaven. Lewis had actually held a contract with the U.S. government since 1812, and Pleasonton renewed it every time it came up. As the number of lighthouses in the country multiplied, every one of them was outfitted with a Lewis lamp.

Lewis, a "tall, fine-looking man of winning address," had captained his own ship by age thirty-five, specializing in the "Liverpool trade" between Boston and England's second most important port. When the United States declared an embargo on trade with England in 1807, during the two countries' squabble over American neutrality in the Napoleonic wars, Lewis made his last trip back from Liverpool late in the season, very nearly wrecking his ship on the dark, gale-stricken coastline. As he recounted to Pleasonton years later, "I could not fetch Province town, or run for Billingsgate island, it being night and no light there. To prevent being driven into Barnstable bay I came to anchor. At day-light, I had lost all my anchors but one. I then ran under the lee of Billingsgate island, came to anchor in 24 feet water, and rode out the gale in safety, by which the lives of a number of passengers and the cargo, valued at 80,000 pounds sterling, in Liverpool, were saved." Safely home but out of a job, Lewis, whose father-in-law Thomas Greenough was a noted "mathematical instrument maker" specializing in nautical navigation, began making lighthouse equipment.

Lewis's system combined mirrored reflectors with a new, improved oil lamp. The design of the lamp was the same as that patented by Aimé Argand in 1780. It had a metal tube placed inside of the wick to allow the flow of air up through the center and another metal tube outside to guide airflow on the wick's exterior. The flame was thus fed not only by the air around the wick (as in earlier lamps), but also by the air within the wick. More air meant greater oxidation, a brighter flame, and less smoke. The addition of a glass chimney on top

The principle behind the Argand lamp,
which Winslow Lewis adopted, was that it directed airflow
both inside and outside a hollow circular wick.

further guided the airflow and increased efficiency. While Lewis did not bring his lamp to the same level of perfection as Argand, it represented an improvement on the spider lamp. Its brightness was equivalent to seven candles, and it burned thirty to forty gallons of oil per year.

Lewis also placed, behind his lamps, reflecting mirrors made of thin copper plated with silver. These could not be called "parabolic reflectors" because he was not terribly careful with the geometry, and the dimensions were often closer to the half spheres that had been long abandoned in Europe. Lewis also placed a glass lens, which he called a "magnifier," in front of the lamp. Measuring nine inches in diameter, it was a good two and a half to four inches thick and set in a copper rim. The idea, of course, was to focus the light rays into a single beam. The arrangement, however, suffered from all of the drawbacks Augustin Fresnel had envisioned for light traversing a single piece of thick glass, as well as the additional drawback that the American-made glass Lewis used was riddled with impurities and had a bottle-green tinge, thus absorbing even more of the light. The effect was to make "a bad light worse," in the words of one inspector.

Lewis claimed both the lamp and the mirror as his original inventions, although he was certainly lying. He insisted that the Argand lamp had never been used in England before 1809 (well after Lewis's last visit there), yet Argand himself moved to England in 1784 to better market his product in the world's industrial hub. Argand had partnered with Matthew Boulton, already famous for the manufacture of the steam engine, to gain as much publicity as possible, and their lamps were widespread by the 1790s. Liverpool, moreover, had some of the most advanced lighthouses of the day. William Hutchinson, the Liverpudlian dock master from 1759 to 1793, had been a technically minded man who experimented with both lamps and mirrors. Indeed, the "Liverpool button" (the addition to lamps of a small disc that helped distribute the inner air current evenly) took its name from the port city. Hutchinson also installed some of the world's earliest and best parabolic mirrors. Coming into the Liverpool harbor, Lewis would have passed the lighthouses at Hoylake and Leasowe, where reflectors had been in use since 1763, as well as Bidston Lighthouse, which from 1771 housed a reflector measuring an astounding thirteen and a half feet across.

Nevertheless, Lewis patented his system in 1810 and set out to sell it. He affixed a lamp to the cupola of the Massachusetts State House and invited all the local authorities, most notably Henry Dearborn, the collector of customs in Boston, to observe his system. The light it gave off was brighter than that from the old spider-lamp system, and, more important, his lamp used about half as much oil. Next, Lewis mounted his apparatus in the Thacher Island Lighthouse near Gloucester, Massachusetts, and then, in May of 1811, at the venerable Boston Light itself. Dearborn, happy with the results, urged Secretary of Treasury Albert Gallatin to buy Lewis's patent.

The following year, Congress voted to offer Lewis an exclusive contract for outfitting the U.S. lighthouses. Lewis pronounced he could put his lamps into all forty-nine of the country's lighthouses for $26,950; in a package deal, Congress gave him $60,000 for the patent, a complete refitting of the lights, and maintenance of them for the next seven years, so long as he could guarantee that his lamps would use half as much oil as the old lamps. Lewis finished forty of the lighthouses by the end of the year, when the War of

1812 intervened, but the work was completed by 1815, and a year later Lewis entered into another contract to keep the lights lit. The federal government provided him with half the amount of oil the previous system had required, plus $1,500 a year for transportation charges and $500 a year to keep the lighting apparatus in repair. This worked out nicely for Lewis, since he was allowed to sell whatever oil was left over at the end, and to keep the profit, which proved to be as high as $12,000 a year. (By comparison, Pleasonton's annual salary was $3,000.) The contract specified that Lewis visit each lighthouse in person to check on the lamp and report back to Congress. He was, in effect, a superintendent of lighthouses for a government that did not have such a position.

Thus, when Pleasonton took over supervision of lighthouses in 1820, now overseeing fifty-five lighthouses, he was more than happy to continue the arrangement with Lewis, the apparent expert. The two men shared an absolute passion for the bottom line, with cost-cutting their top priority. Pleasonton was subject to the government regulation that required officials always to take the low bid; Lewis promised rock-bottom expenditures. Within a year, Pleasonton signed yet another contract with Lewis, who now promised to use just one-third of the oil consumed by the spider lamps. Lewis also began to build lighthouses, becoming the Establishment's most important contractor. By his own estimate, he built some eighty lighthouses over his career. No one, it seemed, could operate a lighthouse as cheaply and efficiently as he could. Pleasonton bragged that he often spent *less* money than Congress had appropriated for him. His careful attention to the bottom line, and Lewis's strict oversight of the construction, meant new lighthouses were built and operated under budget. "The money saved in this manner," Pleasonton attested, "and carried to the surplus fund, is unexampled, so far as I know, in the annals of government."

What is less clear is whether Lewis's lighthouses saved any lives. Ship captains, more concerned with the visibility of the lights than how much money they saved, were unimpressed.

Complaints about Lewis's lights only increased once sailors began to see Fresnel lenses in operation in France. By 1830, several French lighthouses had major seacoast lenses, including those at Cordouan, Dunkirk, Île de Planier, Granville, and Biarritz, and it was hard for the Americans not to notice how much brighter they were. In response, Pleasonton wrote to France inquiring about the lenses. When Léonor Fresnel quoted five thousand dollars for a first-order lens and two thousand for a third-order lens, Pleasonton dismissed the idea. The complaints continued. The dissatisfaction of American sailors found an outlet in the *American Coast Pilot*, a yearly publication providing navigators detailed directions for getting in and out of harbors. Edmund and George W. Blunt, the brothers who published it, did much of the surveying themselves because no government agency took responsibility for it. They meticulously described every U.S. lighthouse, occasionally with excoriating assessments. They asked their readers for tips and soon amassed a small library of seafarers' comments on the dismal state of American lighthouses.

Edmund Blunt toured Europe, visited the lens factories, met Léonor, and returned to campaign strongly for the implementation of the French system. In 1833, the brothers wrote to the secretary of the Treasury about the "great deficiency in our present Light house system" and submitted translations of documents that outlined the details of the Fresnel lens. Failing to get a response, in 1837 they directed their complaints toward Congress. Under Pleasonton, they contended, the lighthouse system lagged far behind those of France and Britain. The lights' distinguishing characteristics were not rationally planned and were liable to change without adequate notification. The towers were shoddily built of poor materials. But worst of all were Winslow Lewis's dreadful lamps, which the Blunts pronounced "nothing but the Argand lamp, with miserable arrangement."

Congress finally took action. In 1838, it halted all money allotted for improvements to the lighthouses and directed the navy to oversee an investigation. When the navy affirmed that the United States should try the Fresnel lens, Congress asked the Committee on Commerce to arrange a test to compare Fresnel lights with the existing reflectors, "by full and satisfactory experiment."

The first Fresnel lens appeared in the United States entirely against Pleasonton's will. Not only did it take a special act of Congress to set up the demonstration, but Pleasonton's response made it clear he was not on board. The lenses themselves made very little difference, he claimed, and any advantage in the French system was a result of their use of the Carcel lamp, an improvement on the Argand lamp. Developed in 1800, the Carcel lamp allowed oil to be stored below the burner and pushed up through the wick by a piston. Pleasonton assured Lewis that the installation of a Fresnel lens did not signal a permanent change in lighthouse policy: "You need be under no apprehension that I mean to leave you out in putting up and repairing our light houses."

Congress intended to have two lenses, one fixed and one revolving, installed in New Jersey on Sandy Hook, the enormous barrier spit stretching deep into New York Bay and causing no end of headaches for ships approaching New

York's busy harbor. Senator John Davis, head of the Committee on Commerce assigned to oversee the trial, noted that this was the most important site in the country for commerce: more sailors passed by it than any other spot. As he put it to Pleasonton, "If the experiment shall be successful they [the sailors] will discuss it. If it fails they will also discuss it and the knowledge in either contingency is precisely what we desire."

Congress assigned the task of acquiring the Fresnel lenses to a young captain, Matthew C. Perry, who had just been given command of the USS *Fulton*, the second steam-powered ship in the navy. Perry was an outspoken advocate of steam in particular and of modernization of the navy in general. He was, moreover, good friends with Edmund Blunt and keenly aware of the lighthouse problem. He was to tour Europe primarily to gather information on the use of steam power, but the secretary of the navy asked him also to investigate the lighthouses of Britain and France, and to buy two Fresnel lenses: a first-order fixed and a second-order revolving lens. He met with Léonor Fresnel and placed the order with Lepaute.

Senator Davis was excited to hear that lighthouses outfitted with Fresnel lenses would, for the same amount of oil, produce three to four times as much light as those with parabolic reflectors. Moreover, Lepaute reported a first-order lens could be seen with the naked eye fifty miles away; a second-order one, forty miles; a third-order lens, twenty-eight miles; and even harbor-lights, twenty miles away (the longer distances were largely theoretical for lighthouse work, since the curvature of the earth imposed an upper limit on how far one could see). In comparison, American lights generally could be seen only ten to sixteen miles away. Davis wrote to Pleasonton, "I should be glad to know your views upon his estimates and especially whether he has allowed the right quantity of oil for the consumption of our lamps. If he has, then he leads us to quite important results." If all were true, Davis marveled, the lighthouses at Sandy Hook would soon have "the best light ever seen." But if he was looking for someone to share his enthusiasm, he had picked the wrong man. Pleasonton remained ill disposed and began an underhanded campaign of sabotage. First, he refused payment for the lenses, although Congress had already appropriated the funds. Léonor Fresnel became concerned; Captain Perry did his best to reassure him, as Perry wrote to a friend, "Mr. Pleasanton has purposely thrown these difficulties in the way. He always strenuously opposed any innovation upon the existing Light House system of our country shamefully defective as it is. . . . Now he will persevere in thwarting the advocates by continuing to interpose delays and objections until the appropriations are exhausted." Perry made clear that the delays emanated from the fifth auditor. "The truth is the old egotist has pronounced his America light house system the best in the world, and was excessively annoyed at the exposure of its utter worthlessness."

Lepaute was eventually paid, and the lenses delivered to New York in March of 1840. But more trouble was in store. Louis Bernard, head of the Lepaute workshop, was delegated to oversee their installation, and given to believe that he should travel to New York for the job. When he disembarked in August, however, he discovered that the U.S. delegation was expecting him in Washington. Finding himself alone in a strange city and unable to speak English, Bernard wrote to the Americans, "Having only received very vague instructions, I will admit to you my difficulty in transporting myself

to Wasington [*sic*], not knowing a word of English and believing the seat of my work to be in New York." By the time of
Bernard's arrival, Pleasonton wrote to Davis, it was too late
in the year to begin installation. Bernard ignored Pleasonton's
suggestion to delay the project for a year, but he found the
Sandy Hook Lighthouse unsuitable. The lantern room was too
small for the first-order lens, and the tower was too low to
produce the full range of the light—the curvature of the earth
would cut it off well before it faded in the distance.

The only two lighthouses in the area high enough to prop-
erly display the lenses were located four miles down the coast
on a tall bluff some two hundred feet above sea level. The
Navesink Highlands, as the site was known, also overlooked
New York Harbor, and although it did not sit as far out as
Sandy Hook, its height made it an ideal location. The "twin
lights" sat on the bluff 320 feet from one another. Bernard,
after moving the delicate equipment, assembled the second-
order fixed lens in the northern tower and the first-order
flashing lens in the southern one, finishing in December.

Next, Pleasonton informed Congress that the equipment
required the hiring of skilled engineers as lighthouse keepers,
which would surely send costs soaring. Bernard gave a full
account to Senator Davis of lighthouse operations in France,
pointing out that keepers were usually old mariners. Léonor,
too, wrote to Congress to assure the Americans that "ordinary mechanics or laborers" would do; he provided a detailed
set of instructions to tell the keepers at Navesink everything
they needed to know.

Technically, the lenses were a success. John Davis had
hoped the sailors heading into New York Harbor would talk
about it, and they did, with frank appreciation. When Congress asked officers of the U.S. Navy to compare the lights

at Navesink with those at Cape Henlopen in Delaware, a bit farther down the coast, Commander Thomas R. Gedney's reply was typical: "The lenticular lights at the Navesink so far surpass in brilliancy the lights at Cape Henlopen and Sandy Hook, as scarcely to admit of a comparison." The Blunt brothers finally had something good to say about American lighthouses in the *Coast Pilot*, reporting that the first-order lens, "on the Fresnel plan, . . . is without doubt the best light on the coast of the United States."

Clear technological superiority was not enough for Pleasonton, though, and he, not Congress, remained in control of the normal operations of the Lighthouse Establishment. The total price for the purchase and installation of the two lenses had come in at $18,975.36, and he rejected the efforts as extravagant. The future for the Fresnel lens in America did not look bright.

"Confusion, Extravagance, and Impotence"

Complaints about American lighthouses continued as it became clear that the Navesink experiment was not going to be repeated. In 1842, eighty-two captains of merchant ships demanded "a thorough examination by competent and disinterested persons." Several-dozen insurance underwriters from New York and Boston added their voices to the call. In response, Congress appointed a young civil engineer, Isaiah William Penn Lewis, to study the lighthouse situation in Maine and Massachusetts. Lewis, or "I.W.P.," as he liked to be known, was Winslow Lewis's nephew, not the obvious choice for a disinterested examiner. But I.W.P. had already begun to signal his independence in 1839 when he won a contract to replace the lantern in the Boston Lighthouse. Although

Boston was the seat of his uncle's patent lamp empire, I.W.P. surprisingly ordered an entire reflector system from London, the first foreign-made apparatus in an American lighthouse. The fourteen 21-inch reflectors were expensive, at $131.50 each (Winslow Lewis charged less than $100 for his lamps), but I.W.P. argued that their superior quality was worth it (and ship captains were vocal in their agreement). Pleasonton characteristically refused to fully reimburse him.

After touring the Northeast coast in 1842, I.W.P. sent a report to Congress that bluntly condemned the system Pleasonton and Winslow Lewis had developed. "Instead of order, economy and utility, in the administration of our Lighthouse system," he wrote, "we have confusion, extravagance, and impotence." Winslow Lewis's virtual monopoly for more than thirty years, uncontested and undeserved, perpetuated a system that was bad to begin with and more hopelessly out-of-date every year. The report concluded with the thought that the lighthouses in the United States stood as "236 witnesses" to "the rule of ignorant and incompetent men."

The ensuing argument between the two Lewis men brimmed with Oedipal and generational tensions, but they cast it as another classic drama: science-minded engineer versus experienced man of the sea. The situation echoed the changeover in the Stevenson family in Scotland, where the self-taught older generation was more dubious and the university-trained younger one more enthusiastic. But the Stevenson revolution was a gentle affair: a loving and dutiful son gently nudging his cautious father toward the new technology. For the Lewises, it was all-out war. The nephew's central accusation was that Winslow was "no man of science." His uncle's refusal to introduce the clearly superior Fresnel arrangement could be grounded only in complete ignorance of engineering and the basic principles of optics. It was for this reason, I.W.P. proposed, that engineers should be in charge of lighthouse operations, specifically the U.S. Army's new Corps of Topographical Engineers, which had been created in 1838 to construct federal civil works. The existing system, whereby anyone, even "an idler or an obscure individual," could slap together a shoddy lighthouse and outfit it with Winslow's feeble lamps, had had disastrous results.

Winslow Lewis counterattacked: his nephew was "an inexperienced individual, who never navigated the shores of Cape Cod twice his life, perhaps not once." The topographical engineers who flaunted their degrees and training knew nothing of the life at sea, or the needs of the men on ships ("his whole knowledge has been obtained through books," he complained to the incoming Treasury secretary in 1843). I.W.P.'s report was a "base and scandalous attack" in which "not one statement . . . is founded on fact." Pleasonton dismissed I.W.P. as "malicious & in general confounded & unworthy of belief"; he had fitted only four lighthouses, and these consumed excessive amounts of oil.

The report was a sensation, but it brought little change. Pleasonton remained firmly in charge, and Winslow Lewis essentially the sole provider of the illuminating apparatus. For his efforts, I.W.P. found himself blacklisted, no longer allowed to inspect lighthouses and the defendant in a series of libel suits over the report. In 1843 he left the country to tour the lighthouses of England and then settled for the better part of two years in France, where he developed a close relationship with Léonor Fresnel and absorbed the principles of the new lenses.

When I.W.P. returned to the United States in 1845, he was determined to get Fresnel lenses into American lighthouses.

Bringing examples of the lenses with him, he gave demonstrations in Boston and New York. After dazzling the Merchants and Underwriters Associations of both cities, he encouraged them to petition the secretary of the Treasury, "praying for the establishment of a dioptric light." Two Harvard physics professors inspected his lenses and reported their advantage over reflectors, "obvious and easy to be understood." Senator John Davis, whom I.W.P. had met at Trinity House and had shown around a factory where Fresnel lenses were made in Paris, wrote letters of support to the secretary of the Treasury, on I.W.P.'s request.

In addition, I.W.P. took another track to circumvent Pleasonton. The Corps of Topographical Engineers had been responsible for building a limited number of lighthouses, which were then handed over to the care of the fifth auditor. The idea was to have the two agencies work in harmony with one another, but the engineers answered directly to Congress. I.W.P. saw his chance. He lobbied Congress to place more lighthouses in the hands of the topographical engineers, and in 1847 it charged them with new construction at Sankaty Head, Massachusetts; Brandywine, Delaware; Minot's Ledge, Massachusetts; and Carysfort Reef and Sand Key Reef, Florida—five difficult locations requiring particular engineering skill.

Sankaty Head, on the island of Nantucket, was the first one completed. I.W.P. had singled out the site in his 1842 report, labeling it "a fatal spot upon the coast of Massachusetts, where many a brave heart and many a gallant ship lie buried in one common grave." Lit on February 1, 1850, with a second-order lens from Lepaute, it immediately became the brightest lighthouse in New England, with reports of its light being visible from the mainland, a good forty-one miles away. Sea captains nicknamed it "the rocket light" and "the blazing star." Fishermen claimed that they could bait their hooks by the flash of the light, even on the darkest night. Visitors came to see the impressive new lens; the staircase opening was even widened to allow ladies with hoop skirts into the lantern room.

Two West Point graduates who became army engineers, Hartman Bache and George Gordon Meade, were in charge of construction for the Brandywine Shoal Light in Delaware Bay. They imported the innovative screwpile technique from Britain, which involved corkscrewing a large number of pilings into the ground to anchor the lighthouse in the bay's sandy shoals. Its third-order lens arrived from Paris before the tower was complete, and I.W.P., now living in Philadelphia, set it up for public display at the city's Franklin Institute for the Promotion of the Mechanic Arts until it could be installed. The lighting itself, on October 31, 1850, was a civic

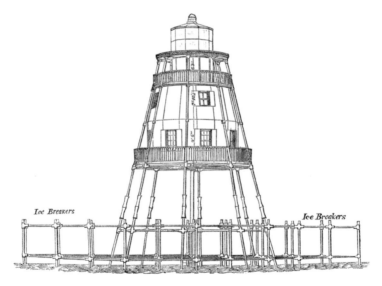

The screwpile lighthouse at Brandywine Shoals.

celebration ("It was an event in those comparatively quiet days," Hartman's nineteen-year-old son later remembered).

The true showcase of modern optical technology was to be two new lighthouses on the long stretch of coral reef that made up the Florida Keys: Sand Key on the western edge and Carysfort on the eastern. I.W.P. took charge of the designs himself, envisioning elegant wrought-iron screwpiles with enormous first-order lenses; the Corps of Topographical Engineers rejected his uncle's bid to build more conventional masonry towers.

Construction began at Carysfort, a difficult spot under four and a half feet of water. Lightships (boats outfitted with lamps) had guarded the reef since 1825, but they themselves kept foundering on the rocks, and sailors who frequented the area thought them so poor as to do more harm than good. As the *Coast Pilot* warned in 1847, "We advise shipmasters not to place much reliance on this, or any of Florida lights, as they are all bad." I.W.P. oversaw the forging of the iron pieces in Philadelphia, where they were assembled to test the towers before they were packed up and shipped down to the Florida Keys. Topographical engineer Captain Howard Stansbury was in charge of construction at the site, although the work was delayed until the arrival of George Gordon Meade, who saw it to completion.

I.W.P. ordered two first-order lenses from Paris and a corrugated iron house, intended as a keeper's dwelling, from England. When news of the purchase found its way to the press, the *Baltimore Sun* raised a cry at the costs and wondered why no American foundries were good enough to make the house. Irate readers complained to the Lighthouse Establishment, whereon Pleasonton seized the opportunity to make it clear that such an act of extravagance would never take place on his watch.

I.W.P. also broke protocol by ordering the lenses directly from Paris, rather than going through the Corps of Topographical Engineers. He humbly accepted the ensuing scolding by the Corps's chief, Colonel John James Abert. But more trouble was to come. Lepaute sent the first lens he completed to the U.S. government rather than the Corps. When it arrived in New York, no one thought to inform the Topographical Engineers, so the lens sat at the Custom House waiting for someone to pay the excise tax.

No one knows why—it was either a case of startling incompetence or a last bit of Pleasonton's devilry—but the lens remained uncollected at the Custom House for the next nine months. Then custom officials followed their standard policy and put the thirteen boxes, contents unidentified, up for auction under the label "machinery." A warehouse employee, a German named Lutz, put in a bid on the assumption that such heavy machinery was bound to be valuable. Winning the set for five hundred dollars, he moved the boxes to a nearby lot and studied the perfectly incomprehensible pieces. Lepaute, meanwhile, had not been paid. Incensed, he ceased work on the Sand Key lens, his last remaining contract with the United States, and he let it be known he would not make another lens for the country.

I.W.P. still had a lighthouse to fill. The tower at Carysfort was an engineering gem, rising magically out of the water. But the real point of it, for I.W.P., was the opportunity to showcase the Fresnel lens, proving at long last the superiority of the Fresnel system. The lens was going to make his lighthouse not just good, but the best in the country. Alas, I.W.P. had to order Meade to equip the tower with eighteen 21-inch-diameter Winslow Lewis patent lamps and reflectors instead.

A Change in the Air

But the winds of change were blowing in the 1840s, ushering in a movement embracing the possibilities of progress and technology. Ralph Waldo Emerson, speaking before Boston's Mercantile Library Association in 1844, coined the term "Young America" to describe the new spirit of the age, entwining modern technology with the expansion into what he called America's "ample geography." This was both a literary and a political movement, fueled by writers such as Walt Whitman and Herman Melville trying to work out what a genuinely American literature would look like. Both men, incidentally, had a keen interest in the technology and power of lighthouses: Melville made it a point to visit the brand new Fresnel lens installed at Sankaty in 1852, while Whitman included a paean to the Montauk Point Lighthouse and its rugged setting in his *Leaves of Grass*.

The movement transformed the American political landscape in the 1840s and 1850s, knitting together a constellation of variously modern causes: scientific and technological progress, increased trade, and government involvement in internal improvements. Behind it all was the pressing issue of territorial expansion. Railroads and telegraphs pulled the country together and made a vast nation spanning the continent conceivable. Steam power allowed more efficient navigation of the Mississippi and the Atlantic. The government's investment in rail, steam, and telegraphs was all done with an eye toward improving trade. The Fresnel lens sat perfectly within this constellation, as an exemplar of scientific technology, an enabler of increased trade, and a compelling argument for government investment.

The first president of this new generation was James K. Polk, who had won the presidency on a single issue: the promise to push the boundaries of the United States as far west as he could. At forty-nine, he was the youngest man to have been elected president, and everything about him seemed modern and forward-looking. Polk's nomination was the first to be transmitted over telegraph, and he was the first president to be photographed in the White House. When France underwent its own revolution in 1848, he welcomed with open arms its provisional government, headed by Fresnel's old mentor, François Arago. But most significantly, he oversaw the dizzying extension of American soil through the annexation of Texas, the Mexican-American War, and the Treaty of Oregon.

One of the first orders of business of his new administration was to take up the lighthouse question. Polk had appointed Robert J. Walker, an equally fervent expansionist, as the secretary of the Treasury. Within weeks of taking office, Walker had sent a delegation of two junior navy officers, Thornton A. Jenkins and Richard Bache Jr., to spend a year in Paris procuring information on the Fresnel lens. After their return in 1846, they described the French system, emphasizing not only the important role of the lens but also the prominence given to scientists and engineers in the administration. Walker forwarded their report to Congress, pointing out that this was exactly the kind of reform the United States needed.

Theirs were not the only voices calling for change. Indeed, Richard Bache Jr. came from a family that produced a remarkable crop of Fresnel supporters whose next generation included Hartman Bache, with the Topographical Engineers; Alexander Dallas Bache, with the United States Coast Survey; and naval officer George M. Bache. They were, moreover, all direct descendants of Benjamin Franklin, himself an ardent champion of lighthouses who saw in them a perfect blend of utility and humanitarianism. Sharing their great-grandfather's fondness for both science and the public good, they all began pushing for lenses in the 1840s and spoke longingly of the day when their own government would heed their advice.

They were joined by Joseph Henry, the first American physicist to rise above amateur status. He had taught himself the new French physics by reading English translations he found in a library, and after discovering the phenomenon of electromagnetic induction independently of Faraday, he became the only homegrown American physicist of his day to be taken seriously on the international stage. Hired by Princeton University, he set out for Paris in 1837 to look for equipment for a first-rate physics laboratory he was putting together from scratch. Henry, abroad for the first time, found himself a bit bewildered by the stylish Parisians, whom he described as "inhabitants almost of an other planet," and admitted being shocked by the naked statues in the Jardin des Tuileries. Most of his time, however, was spent in the workshops of Paris's acclaimed instrument makers, and in particular that of François Soleil (from whom he ultimately ordered nearly a third of the instruments for Princeton's physics lab). One evening in June, Soleil invited Henry for a meeting with Léonor Fresnel, warning him to be prompt since Fresnel was quite punctual and would be lighting his four first-order lenses precisely at 8:00 P.M. Henry, who made it in time, found Léonor intelligent and gentlemanly, but he was above all charmed with the ingenious design. "The effect is very brilliant," he wrote back home, "and being produced by a very simple but philosophical arrangement gave me much pleasure."

Henry soon had a chance to view the philosophical arrangement in action. After leaving Paris, he continued on to Britain to tour the scientific establishments there. Only a few days after his arrival in Edinburgh, Léonor showed up to inspect Scotland's recently introduced Fresnel lenses. The Northern Lighthouse Board had arranged for a steamboat to convey him to the Bell Rock Lighthouse, and they invited Henry to join them for the tour. Dining afterward at the lighthouse of the Isle of May, home of one of Scotland's first Fresnel lenses, the members of the party were called on to give toasts, and Henry, when his turn came, gave a laudatory sketch of Augustin Fresnel and his work. (Presumably, he did not repeat the praise a few days later, when he met with David Brewster.) At the return of his trip, Henry reported with some amazement how, in Europe, the opinions of men of science were taken seriously by policymakers in matters of industry. He hoped that the United States would one day follow the example.

CREATION OF THE LIGHTHOUSE BOARD

Henry's vision for the prominent role of scientific expertise was motivated by the lighthouse engineers he had seen in France, and when the U.S. Congress finally turned its attention to lighthouse reform, Napoléon's Commission des phares was clearly the inspiration. In March of 1851, Congress directed Treasury Secretary Thomas Corwin to create the United States Lighthouse Board consisting of two officers of the navy, two army engineers (the American equivalent to France's Ponts et chaussées), and a civil scientist. A junior naval officer, not a full member of the Board, acted as secretary. The intent was clear: scientists and engineers were in charge, and they were directed to uncover all that was wrong with the existing Lighthouse Establishment.

Pleasonton recognized the attack and immediately fired off an objection. The "junior officer," he pointed out, was none other than Thornton Jenkins, who had a hand in the critical 1846 report, so full of "great injustices" and "strong prejudices." Yet young Jenkins was the very least of Pleasonton's problems. The Board was packed with longtime supporters of the Fresnel system, particularly the engineers. Corwin had asked men with the highest credentials to serve, including Joseph Gilbert Totten, chief engineer of the army, who had been calling for Fresnel lenses since 1840, and Colonel Abert, chief of the Corps of Topographical Engineers, who had just overseen the installation of lenses at Sankaty and Brandywine. Although Abert declined, declaring himself already overextended, his replacement, topographical engineer Lieutenant Colonel James Kearney, similarly supported the Fresnel system.

The navy men, Commodore William Bradford Shubrick and Captain Samuel Du Pont, were also gunning for reform. Although of strikingly different deportment (Shubrick was a man's man, incessantly remembered as "hearty" and "manly," while Du Pont, a member of one of the wealthiest families in the country, was known for his graceful, aristocratic bearing and exquisite taste in interior decorating), both had been hailed as heroes in the recent Mexican-American War and were part of a rising generation keen on modernizing the navy.

The civil scientist was Professor Alexander Dallas Bache, head of the Coast Survey. Bache was, with his close friend Joseph Henry, the most prominent figure in American science and a tireless force for its professionalization. He had taken over the Coast Survey in 1843, transforming it from an insignificant, badly run operation that still relied on the Blunts' *Coast Pilot* to do the hard work of surveying, into one of the country's most important scientific organizations. Bache's model for the transformation was François Arago's Bureau des longitudes, and Arago, watching from France, was one of his greatest supporters. The two had met in 1836–1837, when Bache was conducting a two-year tour of Europe to evaluate methods of scientific education. They hit it off well, making magnetic observations together in the garden of the Observatoire de Paris, and remained in close contact after Bache returned home. Bache stood, Arago claimed, higher than the U.S. naval officers "by a hundred cubits in matters of geodesy and physics of the globe." The admiration was more than returned, as Bache was known to keep a bust of Arago displayed in his office.

This, then, was the secretary of Treasury's crack investigative team. The combination of prestige, technical training, and brain power now focused on the issue of lighthouses was unprecedented. But more than that, all five members represented a wave of reformist impulse intent on dragging the Lighthouse Establishment into the modern world. And unlike the equally impressive group Napoléon had assembled in 1811, this group set to work immediately. "You have seen by the papers that I have been turned over to the Treasury Department," Shubrick, the head of the Board, wrote to his childhood friend James Fenimore Cooper, "this you will perceive is not a small business."

Shubrick's first step was to send Pleasonton a detailed questionnaire, asking pointedly about the number of Fresnel lenses in use and the decisions surrounding installation. Unsurprisingly, Pleasonton was not particularly forthcoming. He rather peevishly replied that the information was already available, and concluded his terse response with the statement "Having heretofore shown that the cost of maintaining our light-houses annually does not exceed one-third of the English . . . I refrain from saying anything further upon the subject at this time." But his heel-dragging would have little effect. The members of the Board had already fanned out along the shores of the coast and Great Lakes, visiting every lighthouse district to gauge the situation for themselves. They toured some forty lights, in addition to looking at many more from the water. To make it fair, they wanted to see the very best reflectors available in this country. Yet even these, they reported, were uniformly awful, with most of them more spherical than parabolic in shape, and many with the silver half worn off.

The Board members were particularly interested in comparing the Fresnel system to the reflectors, and they wanted to do it as scientifically as possible. They borrowed two concepts, developed by Léonor Fresnel, to make their work quantitative: the "useful effect" and the "economical effect" of the lights. Before, the standard way to compare the brightness of two lights was to determine the farthest distance at which they could be seen. This system had obvious problems. Not only did the measurements vary with weather conditions, eyesight, and so on, but the new lights were so bright that the farthest distance at which they could be seen depended on the cutoff of the horizon, rather than their intensity. Léonor thus turned to the science of photometry, adopting some of the techniques Arago had recently worked out during his study of optics. The human eye, scientists had determined, was bad at gauging brightness on any absolute scale, but it was excellent at comparing the brightness of two objects placed next to one another. Léonor arranged a setup, at his control workshop, where he could compare any lens arrangement to a standard lamp by moving the former back and forth until the two lights appeared equally bright to the eye. The distance between them then provided a quantitative measure of brightness. For rotating lenses, Léonor wanted a sense of the total output of light, and not just the brightness of a single beam. So he measured the brightness of the test lens at successive times while it rotated on its clockwork, and added the measurements to obtain a value for the entire rotation. This provided the "useful effect," defined as the total quantity of light projected on the horizon. The "economical effect" was simply the ratio of the useful effect to the annual cost of maintaining the lighthouse. Years of testing had given the French a firm, quantitative argument for how much better dioptric lenses were, and they happily shared the data with the American Lighthouse Board.

Shubrick and the Board followed Léonor's techniques while comparing the lights at Navesink and Sandy Hook. This was supposed to be a comparison of the best lights available, since the twin towers at Navesink had two of the country's few Fresnel lenses and Sandy Hook was one of the rare American lighthouse equipped with the large, twenty-one-inch reflectors (whose expense Pleasonton generally avoided by accepting bids calling for smaller, cheaper reflectors). But on touring the sites, the Board found little to impress them. Sandy Hook's mirrors were out of adjustment and badly scratched, while a decade of use had left Navesink's revolving first-order light "greatly neglected" and out of alignment. The Board was shocked to find that the second-order fixed light at Navesink was actually brighter than the first-order revolving, one, which could only be due to poor arrangement of the prisms. Yet even if the Navesink lamp was not representative of what a first-order revolving light should be, it was still quite a bit brighter than the light at Sandy Hook. The Board members established this quantitatively by going out on the water and moving their boat around until the Sandy Hook and revolving Navesink lights were of equal intensity. (The examiners viewed the lights through colored glass, which diminished the intensity and allowed them to get closer to the lights.) Given that the boat was finally positioned 2¼ miles from Sandy Hook and 5⅞ miles from Navesink, the members calculated the "relative useful effect" of Navesink's light as 5.2 times greater than that of Sandy Hook's. Since the Sandy Hook light was also 1.72 times more expensive, the Board declared its findings "an unanswerable argument in favor of the lens system."

The Brandywine Shoal Lighthouse was also an excellent candidate for comparison, as it sat near two of the country's best Lewis lights, at Cape May in New Jersey and Cape Henlopen in Delaware. Although Brandywine had only a third-order lens, it had the benefit of being brand new and in good condition—a "beautiful specimen of mechanism," the Board proclaimed, without a single scratch, fracture, or imperfect prism. Its light far outshone the May and Henlopen lights, even though the latter were intended as first-class seacoast lights and Brandywine sat well inside Delaware Bay. The Board crunched the numbers, calculating the useful effect of the light from the Fresnel lens at Brandywine to be 5.75 times greater than that at Henlopen and 3 times better than the light at Cape May. The economical effect was even starker, as Henlopen used 3.93 times as much oil as the Brandywine light and Cape May 3.22 times as much.

After performing its comparisons, the Board decided to set the ratio of the economic effect of reflectors to lenses at

3.6 to 1—that is, they claimed the reflector system was 3.6 times as expensive as the lenses. This was an exceptionally conservative estimate, since it did not take into account the gains in brightness the lenses would bring. But the Board wanted to be scrupulously fair, and had little doubt that the numbers would come out in its favor. The next step was to go through the entire list of lights in the United States and estimate the costs of replacing the reflectors with lenses. For example, at Sandy Hook, installing a second-order lens would save 336 gallons of oil, or $459. A lens, moreover, would give off light that was six and a half times brighter, and it would never wear out or need to be replaced, unlike reflectors, which had to be changed every ten to fifteen years. Since lenses needed only a single lamp, instead of the eight to eighteen required by the arrays of reflectors, the Board also estimated a savings in wicks, glass chimneys, and lamp repair at a ratio of 10 to 1. If the entire coast were equipped with Fresnel lenses, the Board calculated, the government would save $112,185.27 a year in oil, supplies, and transportation. Of course, the lenses themselves would be expensive. The Board estimated the cost at just under $500,000, but it figured that at least $90,000 could be recouped by reusing the old reflectors in lightships. And with a savings of well over $100,000 a year, it figured that by the fifth year, the government would see a net gain of over $150,000.

With the Board members busy evaluating the country's lights, its secretary, Thornton Jenkins, sent letters to the European authorities he had met during the fact-gathering trip he and Richard Bache Jr. had taken several years earlier. Jenkins wanted news of the latest improvements. By this time, though, Léonor Fresnel, Jenkins's primary contact, had retired from the Commission des phares, although he still cheerfully responded with what information he had. He also forwarded the questionnaire to his replacement, Léonce Reynaud, alerting him that he was about to become involved in an affair "which appears to me to be of a very serious business." Meanwhile, Thomas Stevenson replied from Scotland, giving more details on his holophotal system, as did Jacob Herbert for Trinity House, William Lord for the port of Liverpool, and Alexander Mitchell, inventor of the screwpile design. Their responses further illustrated just how far behind the United States was. France, for example, not only had installed Fresnel lenses in every single major light on its coast, but was already replacing them with a second generation of improved lights.

America's backwardness was more than confirmed by the responses of her sailors to the questionnaire that Jenkins circulated among the navy and the merchant marine. Most leapt at the chance to respond. David D. Porter, lieutenant in the navy, returned his questionnaire "with great pleasure, as our light-houses as at present arranged are so wretched, that any sea-faring man must desire a change." Letter after letter poured in affirming that lighthouses all over the world were brighter and better arranged than those in America. Many of the letters included the phrase "with the exception of Navesink" in their condemnation, but even these admitted that the New Jersey lights, by far the best on the coast, were unimpressive by European standards. Typical was the complaint of an officer aboard the *Illinois* about the lights of Hatteras, Lookout, Canaveral, and Cape Florida: "they had better be dispensed with, as the navigator is apt to run ashore looking for them." J. C. Delano concluded his letter with a sailor's forthrightness: "I do hope you will be able to show to a committee of Congress such a ****** [*sic*] state of things, that that body will no longer withhold the means of improvement."

After nearly a year of compiling complaints, touring the coasts, and comparing lights, the Board issued its report, a massive tome of 760 pages. It had both a specific argument and a broader theme. The argument was that Pleasonton's rejection of the Fresnel lens led to a terrible degree of waste, inefficiency, and unnecessary danger. The Board backed up its claim with page after page of calculations showing how easy it would be to get more light for less oil. The broader theme proved just as urgent: the American lighthouse service needed to modernize and learn some science. The most notable physicists of France, Scotland, and England had been deeply involved in their country's lighthouse establishments, marking a fundamentally different attitude about the importance of science. Pleasonton, the report pointed out, may or may not have been a fine accountant, but he was in no position to control a lighthouse system he did not have the capacity to understand. The proper planning for lighthouses, they wrote, "is the business of an engineer." The unscientific nature of American lighthouses was an embarrassment, and to expect anything else from the current system would be to expect "order out of anarchy and confusion." What was needed was a rational plan along the lines of the French model, with top scientists in charge and trained engineers responsible for all of the building.

The report was a frank call for immediate change. But still Congress dithered, consumed by the politics of such a drastic overhaul. The Fresnel advocates grew impatient, and George Blunt openly dreamed of sticking the whole of Congress on a boat off a rocky shore during a storm and letting the elected representatives see for themselves how bad the situation was: "Could I but put them on a topsail yard on a stormy, wintry night, on a lee shore, to lookout for our sickly, flickering uncertainties, I have no doubt that when snug below, the 'Old Fogies,' 'Young America,' Webster, Fillmore and Scott men would all unite in crying for more light." The stalemate between tradition and reform, or the Old Fogies and Young America, continued for months. Then, in the fall of 1852, Blunt got his wish. As political reporter George Alfred Townsend recounted,

> In 1852, the bill for creating the Light-House Board was pending in Congress, but being opposed by parties interested in keeping up our bad system, its passage was doubtful. The Baltic steamer was then at Washington and sailed to New York. Off Sandy Hook, she was detained by a fog, and could not run for want of proper buoys. A meeting of the passengers, among whom were several members of Congress, was called on board, and their attention particularly directed to this defect, and on their returning to Washington, they caused the above-named bill to be passed.

After suffering firsthand the deficiencies of the Sandy Hook reflectors, Congress finally, on October 9, 1852, passed the law placing care of the country's lighthouses under the new, independent Lighthouse Board. This permanent Board shared both the name and membership of its temporary predecessor, with one significant addition: Joseph Henry joined as a second civil scientist, chairing the Committee on Experiments modeled on Augustin Fresnel's control workshop. Henry had been appointed the first director of the Smithsonian Institution in 1846, and now he made room in the Castle on the National Mall for the necessary experiments, setting up a photometric lab to compare the brightness of various apparatus and arrangements. He, with the rest of the Board, focused his attention on the Fresnel lens, which would clearly be the centerpiece of the reform to come.

RELIGHTING THE COAST

The Board's task was enormous: building, in its own words, "a uniform and systematic plan upon the ruins of the old and unsystematic establishment." The Board started by dividing the coast into twelve lighthouse districts, each supervised by an officer of the army or navy (replacing the overworked customs collectors who were in charge under Pleasonton's watch). The Board then sat down for the complicated task of deciding which lenses would go where. Inspired by the French "rational system" of trying to stagger the lights so that no part of the coast would be left dark, the members also took into account the peculiarities and sheer expanse of the American coastline. Their wish list ultimately included over three hundred lenses, almost twice as many as France had installed, and it still left a few dark gaps. At the top of the list was the single most troubling spot on American shores: Cape Hatteras, the infamous "graveyard of the Atlantic" that had taken down more ships than any other stretch along the coast. Every ship going up or down the coast passed it, and the Gulf Stream ran so close to the shore that most sailors wanted to cut as close to land as possible. There had been a lighthouse on Hatteras since 1803, but it was, in the words of the Blunt brothers, "notoriously bad." Worse than useless, it was downright dangerous. The shoals extended more than ten miles out into the sea, and while the light from the reflectors could in theory be seen fourteen miles away, in practice it almost never was. The lighthouse was thus little more than a decoy luring ships onto the shoals as they searched for a light too dim to see. Experienced sailors knew to ignore the light entirely, with virtually every one of the nautical men the Board interviewed begging the Board to replace it, "the most important on our coast, and, without doubt, the *worst* light in the world."

Before anything could do done, however, the United States needed to repair its relations with the lens manufacturers, left tattered by the Carysfort debacle. Augustin Lepaute was one of the authorities Jenkins had written asking about his firm's latest improvements. Lepaute wryly responded that his company had just made a beautiful first-order lens using a new, improved plan. He had displayed it at the 1849 Exposition nationale (France's national industrial exhibition), where it won a gold medal and universal acclaim. He then shipped it to the United States, where it was never heard of again (and remained unpaid for). Another lens of the improved design, intended for Sand Key, had been sitting for months with the work order stopped.

Treasury Secretary Corwin was well aware of the problem. Even before the bill creating the Lighthouse Board passed Congress, he sent naval officer Washington A. Bartlett as a "special agent of the Treasury Department in Paris" to mend relations. Bartlett tried to persuade the French of the massive change in American attitude, and he eventually convinced Lepaute to resume work on the Sand Key lens. Corwin next authorized Bartlett to order another eight lenses, all destined for the California coast. (This was most likely the reason Corwin had chosen Bartlett in the first place, as he had been part of the crew surveying the Pacific coast in 1849–1850, and had been outspoken about the need for good lights in San Francisco's harbor.)

Meanwhile, I.W.P. Lewis had tracked down the missing Carysfort lens, and in November of 1852, he asked the Lighthouse Board to pay for its purchase. Jenkins responded that the Board would need to examine it first, and requested that it be sent to Joseph Henry at the Smithsonian. There it joined several of the other lenses Henry had been testing, but as the only first-order lens among them, its ten-foot frame caused a sensation when it was put on display. (The nearby Metropolitan Mechanics Institute requested to borrow it so that it, too, could entice the crowds with a public display.) Satisfied with the lens, the Lighthouse Board paid for it and its rotating mechanism, with Lepaute recouping six thousand of the nine thousand dollars he had hoped to get for the apparatus. I.W.P. began preparing the Sand Key Lighthouse, his second screwpile tower on the Florida Keys, and the destination for the Carysfort lens, for installation.

With the apparatus originally intended for Sand Key receiving its finishing touches in Lepaute's workshop, the Board was now free to reassign it to the high-priority Cape Hatteras Lighthouse, which it promptly did. But first, the lens had another task: a star turn in New York, at the Exhibition of the Industry of All Nations in 1853. As the American response to the 1851 Great Exhibition in London, the New York affair featured its own Crystal Palace of iron and glass and the very best of American technology. Samuel Du Pont, the general superintendent of the event, made sure the lens would be there for the show. He brought George Gordon Meade, the engineer who had just finished installing the first-order lens at Sand Key, up from Florida to help assemble the apparatus, whose 1,008 prisms made up an astonishing twenty-four sides. Located at the entrance to the South Nave in the Crystal Palace, the lens was impossible to miss, particularly once

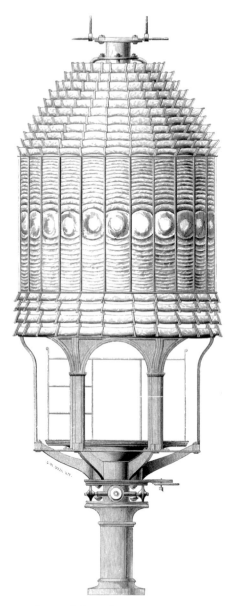

Lepaute's twenty-four-sided revolving apparatus, originally ordered for the lighthouse at Sand Key, then reassigned to Cape Hatteras.

the decision was made to keep the exhibit open at night, to allow people to see the light shining in the dark. Meade marveled at the quick bursts of light produced by the many sides: "Thus, this marvellous contrivance darts forth its dazzling flash, and revolving as it flashes, only intermits its light still more to startle the beholder."

The Board was convinced the spectacular lens would produce the "best results upon the public mind" in favor of the immense transformation it was about to undertake. As the exhibition opened in New York, Jenkins began sending to France large orders for new lenses, dividing the work between Lepaute and Sautter with almost mathematical precision. (Bartlett, who was still in Paris, at first claimed he was entitled to a three percent commission on the orders, which would eventually add up to some half a million dollars, although he was soon found guilty of falsely submitting receipts for reimbursement and discharged from the navy.)

By this time, the lens manufacturers were using a six-part size division. First-order lights were still the biggest, with 6-foot diameters. The other major lights, the second and third orders, had diameters of 4 feet 7 inches and 3 feet $3\frac{3}{8}$ inches, respectively. The minor lights, intended for more local hazards, were now divided into fourth and fifth orders ($19\frac{5}{8}$ inches and $14\frac{1}{2}$ inches, respectively), and the sixth order ($11\frac{3}{4}$ inches) comprised the harbor light. The Board wanted a lot of every size but acknowledged that there was no way to do this all at once. As Jenkins put it in his response to one impatient seafarer, "The Board hopes to be able to remedy all the evils existing in the Light House Establishment, but it cannot be done in a day."

The first request concentrated primarily on the smaller lenses. There was a certain logic to this. Although the big seacoast lights would save more lives, it was the minor lights that would save more money. A sixth-order lens might be no bigger than a gallon jug and cost only $216, but it could save almost $500 a year on oil and pay for itself in six months. With the larger lenses, however, the prices grew exponentially. A third-order lens cost $1,860; a second-order lens, $4,400; and a first-order one, $6,800. (And this was the cost of just the bare-bones lens—the Hattaras apparatus, with its rotating arrangement, wound up costing over $9,000.) Furthermore, it was not easy to say how much oil a first- or even second-order lens might save when compared to a reflector, because there *were* no comparable reflectors. A first-order lens might very well use just as much oil as whatever device it replaced, with the primary result being a vastly brighter light.

Both lensmakers began working full time for the Americans, but the process still took time. Lens manufacture had become a big business in the past thirty years, but it was far from an automated process. Grinding, polishing, and assembling remained painstaking work, and Sautter and Lepaute promised only that the United States would get the lenses when it got them. A few minor lights began to trickle in by October, but mostly the orders remained unfilled in the spring of 1854. Jenkins fired off several increasingly stern letters reminding the firms that the season for lighthouse renovations was upon them, and the United States needed lenses to complete the work. He tried both the carrot ("If the present orders are filled during the present spring and summer, a large order may be expected in the month of September or October next") and the stick ("Soon, we shall be reduced to the necessity of providing reflectors for the new light houses for which these lenses were ordered"). He begged both firms to send "any and all" lower-order lights they had available, regardless of what had been ordered.

By the middle of the summer, the first order was largely complete, and Congress began to release more money for the purchase of big, seacoast lenses. Slowly, every lighthouse on the Board's wish list would be outfitted. Every few months, Jenkins sent off new orders to Paris, interspersed with constant entreaties to fill them as fast as possible. At some points, emotions ran high. On August 7, 1855, Jenkins wrote, "If Messrs Sautter & Co are attending to European orders received subsequent to those from us which seemed very likely, it is very unjust to us, and will not benefit them." Occasionally, after waiting over a year, the Lighthouse Board received a lens with some defective prisms. This was not particularly unusual: Augustin Fresnel and Alan Stevenson had constantly battled with finding pieces of glass that were not up to par. But the time and distances involved rendered the problem especially galling, and Jenkins frequently admonished his manufacturers (and particularly Sautter) to carefully inspect every piece before sending it over. Some delays were not entirely unwelcome. In February of 1855, Lepaute asked that he hold on to one of the lenses a bit longer so that he could display it at the 1855 Exposition universelle in Paris. The United States, sensing some good publicity, assented, with the stipulation that Lepaute specify he had built it for the U.S. Lighthouse Board.

The Board prioritized which lighthouses would get their lenses first, taking into account the site's "importance to commerce" and the danger of the existing situation. Some, such as the Cape Cod and Montauk Point Lighthouses, were important colonial lights badly in need of improvement. Some, such as at Absecon, New Jersey, and Great West Bay, New York, were new lighthouses intended to fill any gaps in the Eastern Seaboard, while others (at Point Conception, California, and Cape Flattery, Washington) were new light-houses illuminating a few spots along the vast darkness of the Pacific coast. Occasionally, the Board got philosophical. When discussing the situation at Cape Henlopen, Delaware, Board members hinted at the delicate balance they sought to strike between preserving assets and saving lives when they wrote, "The large amount of trade from Philadelphia war-rants the proposed expenditure, and humanity would seem to dictate it as consistent with true policy and philanthropy."

PACIFIC COAST

The Board's task was not made any easier by the fact that the United States had more than doubled its coastline in the previous few years, with the addition of Florida and Texas as states in 1845, the Oregon Territory in 1846, and California in 1848. The members of the Lighthouse Board had been enthusiastic supporters of the country's rapid expansion, and lighting the new coastline was a natural step for solidifying strategic and political control of the areas. The discovery of gold in California in 1848 gave these lighthouses tremendous economic urgency as well, as San Francisco was transformed from a sleepy backwater to one of the nation's busiest har-bors, behind only New York, Boston, and New Orleans in volume of foreign commerce.

California became a particular problem for the Board. As soon as the United States took possession of the land, A. D. Bache sent the Coast Survey to map the new coastline and establish the most suitable sites for lighthouses. Based on the agency's work, in 1850 Congress apportioned generous funds for eight new lighthouses on the Pacific. Bache was very specific about which kinds of lights should go where, and that all of these important lights should have Fresnel lenses. But Stephen Pleasonton was still in charge of the Lighthouse Establishment in 1850, and he ignored Bache's recommendations. He proceeded, instead, precisely as he always had: accepting contracts for the job and going with the cheapest one. Winslow Lewis, as usual, supplied his patented lamps and reflectors. Construction of the lighthouses themselves fell, under somewhat suspicious circumstances, to the Baltimore firm of Francis A. Gibbons and Francis X. Kelly, who set out in mid-1852 on a barque, the *Oriole*, laden with supplies. The trip around Cape Horn took anywhere from five to eight months during this era, so by the time they arrived in San Francisco, in December of 1852, the new Lighthouse Board was in place, and their original boss, Pleasonton, was no longer in charge. With no way of knowing, however, they started work in San Francisco Bay, building towers at Alcatraz, Fort Point, Point Pinos, and the Farallon Islands. Still not aware of the changes back east, in the fall of 1853, they finished up and headed north to Cape Disappointment in the Oregon Territory. After their ship crashed on the very shoals they had been sent to light, losing its entire cargo, Gibbons & Kelly ordered yet another set of reflectors to be sent around Cape Horn, despite the immediate order the Board had already issued to cease all construction with reflectors.

Shubrick fumed at the Gibbons & Kelly contract, calling it nothing more than "a fruitful source of difficulty, delay, and expense." The Board immediately issued orders to leave all the Lewis reflectors packed in their boxes while it rushed to order Fresnel lenses from France. But beyond that, it was unhappy that these, the first lighthouses of America's vast new frontier, were in the hands of Pleasonton's dubious vendors. The situation was not improved by the rapid deterioration of Gibbons's previous lighthouses at Bodie Island off North Carolina and Egmont Key in Florida, both of which now needed replacing.

The Board wanted a trained engineer out west, and they chose Hartman Bache, who had been promoted to major in 1838 and now ranked behind only John James Abert and James Kearney in the Corps of Topographical Engineers. Delayed by some projects on the Atlantic coast and a "sudden and protracted illness" likely connected to the 1853 yellow fever epidemic, Hartman didn't arrive in California until 1855, when he took over operations from Captain Campbell Graham, who had been serving as the lighthouse inspector of the twelfth district covering the entire Pacific coast.

When Bache sailed into San Francisco Bay on June 30, 1855, three Fresnel lenses greeted him: the harbor lights of Alcatraz, Fort Point, and Point Bonita. The lighthouse on Alcatraz, a tiny island that ships confronted after passing

Hartman Bache sketched the Pacific coast lighthouses as he oversaw their completion. Alcatraz, shown here, was already lit by the time he arrived.

*Map of California, showing the sites of lighthouse construction
at Alcatraz, Fort Point, Point Bonita, Southeast Farallon Island,
Point Pinos, Point Conception, and Point Loma.*

through the Golden Gate straits, had been the first to go up, on June 1, 1854. Gibbons & Kelly had completed the structure early on, in July of 1853, but it had to wait almost another year while the Board rushed its lens to California as fast as it could. The third-order fixed lens was the first major one to arrive from the Board's long shopping lists, and the first lens from Sautter to be used in the United States.

Fort Point had been the second lighthouse Gibbons & Kelly completed, but that structure had already been razed by the time Bache arrived. The army had decided it wanted to build a fort on the lighthouse's strategic spot on the southern flank of the narrow entrance to San Francisco Bay, so three months after its completion the army tore down the Gibbons & Kelly tower and replaced it with a smaller structure of its own in the narrow space between the fort and the water. It was outfitted with a fifth-order light.

Point Bonita (or Boneta, as it was first designated) marked the northern entrance to San Francisco Bay. It was not on the Board's original wish list, but was part of a second wave of Pacific lighthouses that Congress authorized in 1853 and 1854. Graham arranged for local contractors to build it, and with no Gibbons & Kelly structure to tear down, work proceeded quickly. It was first lit on May 2, 1855, a few weeks before Bache arrived.

The harbor lights may have been a nice welcome, but noticeably absent was the city's great seacoast light, intended for the Farallon Islands. This light would be the first thing ships saw as they approached the coast and the most important light of the Pacific. It had been, after Cape Hatteras, the top light on the Lighthouse Board's original priority list. Even before the Board officially existed, Treasury Secretary Corwin had rushed to order a spectacular first-order revolving lens from Lepaute and pressured the company to complete it as soon as it could. Now, however, after the long journey from Marseilles the lens sat in seventy-three unopened crates on the San Francisco docks, waiting for Bache's pronouncement.

Bache set out to see the lighthouse site soon after he arrived, catching a ride on the Coast Survey steamship to Southeast Farallon Island, thirty miles out. The tower essentially complete, Bache pronounced the construction "a tolerably fair job." The problem, however, was that the lantern room could not accommodate a first-order Fresnel lens. Graham had already hinted as much, but he waited for Bache's inspection to confirm that no amount of effort could make it work. The only thing to do was tear down the brand-new lighthouse and start again from scratch.

The prospect was beyond disheartening. Constructing the tower had been a grueling ordeal. The islands were more remote than anything on the Atlantic coast, and had no practical landing sites. To get to the tower site at the top of a hill, one had to scramble several hundred feet up a steep rock face, which for most of the way was at an angle, as Bache pointed out, "never less than 45 degrees" and often up to 65 degrees. Workers scaled the rock with their hands and feet, carrying only four or five bricks at a time on their backs. Just finishing the tower had been an accomplishment, and the Lighthouse Board balked at the idea of doing it all over again.

Yet rebuilding the tower was not even the hard part. If it was done once, Bache pointed out, it could be done again. Installing the lens was what worried him, a difficult task that he admitted would be "only second to an impossible one." Unlike the bricks, which were small enough to be carried, the lens could not be broken into a manageable size. The delicate components had to be transported in their original packaging, and these were huge crates, up to sixty-seven cubic feet (think of a large couch). Some of the packages were exceptionally heavy, some were awkwardly shaped, some were fragile and required special handling. To solve the problem, Bache decided to build a road to the top of the hill, adding another $1,500 to the bill and even more for the expense of a mule and a boat. "The cost in all cases must be large," he frankly admitted, "and to those unaccustomed to the current prices on this coast will appear even more so. The disbursing officer may regret but cannot change the state of things." There was little the Board could do. Its crack engineer had spoken, and he soon had his road, mule, and boat.

After breaking the costly news about the Farallon lighthouse, Bache headed south to tour the other lighthouses currently under construction. Unfortunately, he had only more bad news to give, finding every one of the towers either inadequately built, poorly located, or of inappropriate dimensions (and most often all three). Only one of them, at Point Pinos, survived his inspection, and that perhaps only because he did not have time to get a close enough look. This lighthouse, located at the outermost tip of the Monterey Peninsula, overlooking the city that had until recently been the capital of California, had, by luck, become the second lighthouse lit on the West Coast. Although the Board members had not placed its intended second-order Fresnel lens particularly high on their priority list, they decided it was the best spot for the third-order light that was originally planned for Fort Point, whose tower had already been dismantled by the time its lens had arrived from France.

Bache saw the light in action when his steamship docked in Monterey in August of 1855, although it did not spend enough time in port for him to get to the tower itself. Still, he found a fair bit to complain about. The site, he claimed, worked neither as a coast light nor as a harbor light. The Coast Survey had suggested three different possibilities, and the construction crew, left to their own devices, had chosen the easiest of the three, farther inland than the rest. It was right at the edge of the pine forest that gave the point its name, which meant the bright new light was obscured by trees from a number of angles. Bache left with the promise that he would be back to do a proper survey and improve the situation, a task he apparently accomplished. By the time Robert Louis Stevenson visited in 1879, walking from the nearby town of Carmel, where he was living at the time, he described the lighthouse as sitting in "a wilderness of sand."

Hartman Bache's next stop was Point Conception, a particularly tricky spot around Santa Barbara where A. D. Bache had recommended a first-order seacoast light. Pleasonton, of course, had ordered something else, and Gibbons & Kelly had built a tower too small for a lens with a six-foot diameter. Hartman, familiar with the plans, had already ordered the old tower torn down to the foundation, which had been done by the time he arrived. He stayed four days, making a few minor adjustments, and left as his local builder, Mr. Merrill, lay the first bricks for the new tower. The last stop on the southern tour was San Diego, where Hartman arrived a few weeks after the lighthouse's new lens did. This time he let the tower stand, and after ordering several modifications, he gave the responsibility for installation of the lens to the workmen who had accompanied it.

After San Diego, Hartman headed north to Cape Disappointment on the Columbia River. Again, based on the

Photo of Point Pinos by Eadweard Muybridge.
In 1871, the Lighthouse Board hired the celebrated photographer
to photograph the West Coast lighthouses.

Hartman Bache's sketch of Point Conception, where he ordered the
rebuilding of the lighthouse to accommodate a first-order lens.

original survey, A. D. Bache had insisted that the spot needed a first-order Fresnel lens, "of a power not less than the best light on Navesink." Yet Pleasonton had ignored him, and Gibbons & Kelly had built a structure that could not possibly hold the twelve-foot-tall lens that eventually arrived. With an original appropriation of $53,000, the expense of the Cape Disappointment Lighthouse had already far outstripped the cost of the other Pacific lights (which were generally budgeted for $15,000, itself a remarkable amount). Yet Hartman ordered its demolition, brick by brick, and a rebuilding effort that took two years to complete. Such was the commitment of the Lighthouse Board to the cause of the Fresnel lens.

By 1859, the Lighthouse Board had put a Fresnel lens in virtually every lighthouse in the United States. The task had been mind-boggling, only made more so by the simultaneous burst of construction of new lighthouses. Already by 1856, *Putnam's Magazine* reported, the United States had twice as many lights as the entire British Isles, and nearly one-third as many as all the other nations of the world combined. And all of them now had the best lenses technology could provide. In a few short years, the United States had gone from having the least number of dioptrics of any developed seafaring nation to the most, by a staggering margin. Meeting or surpassing expectations, the lenses, the Lighthouse Board reported, gave more than four times as much light as the very best reflectors, and only consumed one-fourth the amount of oil.

Léonor Fresnel, watching from France, fully appreciated how amazing the transformation had been: "The prodigious development of this service within so short a time under the Light-House Board has truly astonished me. My old experience in fact enables me the better to appreciate how much energy and activity were necessary to bring to this degree of perfection, the light-house service of such a vast expanse of coast." And a vast expanse it was. More than five hundred Fresnel lenses lit the shores of the Atlantic, the Gulf of Mexico, and various lakes. The Pacific coast had twenty-one—not nearly as large a number, but in many ways just as impressive an achievement. They lit, after all, a long stretch of what had been the country's distant frontier, and marked, with their technological conquest of such difficult terrain, a collective sense of destiny.

SIX

EVERYTHING RECKLESSLY BROKEN

WITH FRESNEL LENSES now installed in virtually every lighthouse on the American coastline, the Lighthouse Board could step back a bit. Commodore William Shubrick, having overseen the monumental operation, left on a military mission to Paraguay, accompanied by Thornton Jenkins, the Board's secretary. Replacing Jenkins was Raphael Semmes, a lighthouse inspector serving the Caribbean and Gulf coasts. Appointed in September of 1858, Semmes moved his family, along with his house slaves, from his home in Mobile, Alabama, to Maryland, near Washington, D.C.

Washington was a tense and divided city, as the differences between North and South grew more manifestly irreconcilable. Lincoln's election in 1860 exacerbated the country's polarization, and prompted South Carolina to withdraw from the Union seven weeks later. Although born in Maryland, Semmes thoroughly considered himself a son of Alabama, and after South Carolina seceded, he secretly let the Alabama congressional members know that he would fight with them if Alabama followed South Carolina's example. As he wrote to a friend in February of 1861, "I am still at my post at the Light-House Board, performing my routine duties, but listening with an aching ear and beating heart, for the first sounds of the great disruption which is at hand." He did not have long to wait.

"The Tocsin of War"

By 1860, the lighthouses in Charleston Harbor had several new Fresnel lenses. Most impressive was the revolving first-order lens in the Charleston Lighthouse, a major seacoast light on Morris Island. The lighthouses at the harbor's three defensive forts also had a Fresnel lens: a fifth-order one at Castle Pinckney, a sixth-order lens on Sullivan's Island near Fort Moultrie, and a fourth-order lens at Fort Sumter. Although Fort Sumter was still under construction, its lighthouse had operated since 1855 as part of a range line, where ships lined it up with light on the steeple of St. Philip's Church to follow the narrow channel into the port of Charleston.

When the South Carolina Secession Convention met on December 17, 1860, the delegates decreed all this property theirs and sent envoys to Washington to negotiate its transfer to the state, implying that if the federal government was unwilling to give it up, the state would simply take it. The lighthouse inspector for Charleston, T. T. Hunter of the U.S. Navy, noted with alarm that South Carolina troops stood poised to seize the lighthouses. He wrote to Washington for instructions, but his letter landed on the desk of Secretary Semmes, who urged the secretary of the Treasury to cede the lighthouses and warned that he would not recommend "that the coast of South Carolina be lighted by the Federal Government against her will."

On December 20, 1860, South Carolina formally seceded. Church bells tolled; revelers lit firecrackers and rockets and even fired cannons. The sounds of celebration in Charleston

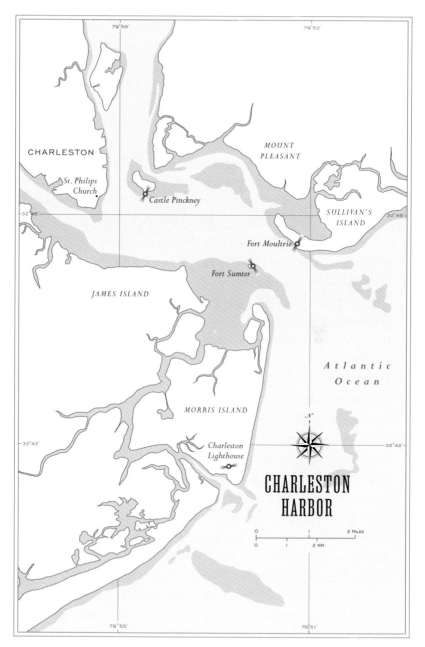

Map of Charleston Harbor.

could be heard at Fort Moultrie, four miles away. The First U.S. Artillery Regiment, garrisoned there under command of Major Robert Anderson, watched nervously as South Carolina troops took up position around Charleston's federal forts. Fort Moultrie's defenses all pointed seaward, and the regiment could not protect the fort from a land attack. Across the channel on an island, however, the more easily defended Fort Sumter was nearing completion. On December 26, under the cover of night, Anderson sneaked his troops into Fort Sumter and prepared to fight. They were joined the next day by the small contingent from Castle Pinckney.

South Carolina's governor, Francis Pickens, worried about the possibility of Washington sending reinforcements, ordered all of Charleston's lights shut down and the buoys pulled up. He directed lighthouse inspector Hunter to leave the state, which he tried to do on a tender (one of the ships used to supply lighthouses), only to be captured. After he surrendered the ship, which Pickens claimed for South Carolina, Hunter was allowed to leave, by land, for Wilmington, North Carolina. Pickens continued to seize federal property, towing the Rattlesnake Shoal lightship into Charleston Harbor and confiscating the remaining lighthouse tenders.

The day after Pickens ordered the lights turned off, steamships arrived on Morris Island and began unloading supplies for the hasty construction of a new fort next to the Charleston Lighthouse. Fort Morris went up quickly, armed with a battery of field howitzers to guard the major shipping channel. The lighthouse tower was converted into a lookout post, manned by cadets from the Citadel, South Carolina's military college, who spent the days scanning the sea for federal ships.

Meanwhile, Fort Sumter was out of coal and running low on food. In Washington, General Winfield Scott prepared to send supplies and reinforcements, leasing a civilian ship, *Star of the West*, for the job. Governor Pickens, well aware of the ship's mission, ordered the troops at both Fort Moultrie and Fort Morris to fire upon it when it approached, and fearing the possible loss of Morris Island, he directed the South Carolina troops to remove the Fresnel lens from the lighthouse and bury it.

The *Star of the West* arrived around 1:30 on the morning of January 9, 1861, and the crew noted that the harbor was dark. Only one faint light, presumably at Fort Sumter, shone. They would have to wait until daybreak to navigate the harbor. At dawn, the cadets on Morris Island spotted the ship and prepared for action. As it drew abreast, they fired a warning shot, then three more rounds. Two shots hit the ship; though neither did serious damage, the ship turned around and headed back to New York, leaving Fort Sumter without reinforcements.

Washington made no further efforts to reinforce Fort Sumter. President James Buchanan, in the last months of his term, still hoped to avoid a war. Neither did Governor Pickens attack the fort, having been warned that he did not have the ammunition to bring it down. A stalemate ensued, while around them Southern states pulled out of the Union. As the secessionists formed themselves into the Confederate States of America, their president, Jefferson Davis, wrote to Pickens telling him that the Confederacy would be willing to take over the problem of Fort Sumter.

Up in Washington, Semmes finally heard the news he was waiting for. On February 14, 1861, the Committee on Naval Affairs of the Provisional Government of the Confederate States of America in Montgomery, Alabama, sent a telegram tersely asking that he "repair to this place at your earliest convenience." Only two days earlier, Semmes had been promoted to a full member of the Lighthouse Board to replace the recently deceased naval officer Edward Tilton (who had replaced Du Pont when the latter was posted to Asia). But he did not hesitate to join the South. The telegram reached him at four o'clock, and Semmes replied that same evening. The next day, he resigned his position, severing all contact with the U.S. government. The day after that, he was on the train to Montgomery. The secessionists had also courted Commodore Shubrick, a South Carolinian, but he refused to leave, much to the scorn of Semmes, who remained the only Board member to join the Confederacy.

Semmes traveled to New York in March to buy munitions for the South. When he returned to Montgomery, Stephen R. Mallory, former U.S. senator turned secretary of the Confederate navy, placed him in charge of the newly created Confederate Lighthouse Bureau. "I had barely time to appoint the necessary clerks and open a set of books," he later recounted, "before Fort Sumter was fired upon, and the tocsin of war was sounded."

Lincoln took office on March 4, 1861, promising in his inaugural address to "possess the property" of the government. He ordered a flotilla down to Charleston on April 4. Jefferson Davis, hearing the news, ordered the attack on the fort. At 4:30 on the morning of April 12, the batteries surrounding the fort began methodically pounding it with artillery. After heavy barrage, the Union troops flew the white flag of surrender.

War had begun. On April 19, 1861, as his first official act of war, Lincoln ordered a blockade of the Southern coast, divided between an Atlantic and a Gulf blockade. The South responded by putting out all the lights in its lighthouses. By April 24, everything was dark. Semmes, sitting in his newly furnished office in Montgomery, wrote that the Lighthouse Bureau was "no longer to be thought of," and every man who could wield a sword should do so. He immediately resigned his position and went looking for a ship, finding a small steamer that had been condemned as unfit and left in New Orleans. A bit of outfitting turned it into the CSS *Sumter*, the first war vessel of the Confederate navy. At its command, Semmes specialized in raiding the unprotected ships of the Northern merchant fleet, eventually capturing eighty-seven and largely driving the fleet from the seas. The North cursed him as a privateer, while the South celebrated his daring exploits slipping through the blockade.

The blockade was intended to keep the South from getting the "arms, munitions, iron plating for armored ships, ship engines and machinery, rail iron, railroad locomotives and cars, boots, cloth, niter, and other manufactured goods" that it had to import, and to cut off the region's chief source of revenue: cotton. The U.S. Navy planned to post steamships around all of the major ports, which could then chase down the vessels that accounted for the bulk of Southern trade. But at the start of the war, the federal navy had only forty-two ships to its name. And once Virginia and North Carolina joined the Confederacy, there were 3,549 miles of coastline to patrol. The blockade thus proceeded slowly, as the Union scrambled to expand the fleet. Confederate runners slipped easily through the blockade in these early months, usually under the cover of night. The local Southern crews knew the hidden hazards of the harbors better than the Northerners trying to chase them down.

Charleston was one of the few harbors to remain in Confederate hands throughout the war, and it became an important base for blockade runners. The lens of the venerable Charleston Lighthouse had already been dismantled and removed by the time hostilities started. As the Union steamships moved into the Southern waters, the Confederates worried they might lose Morris Island, and took steps to ensure that the tower would also be inoperable. According to the *Charleston Mercury*, on December 20, 1861, "The report reached us yesterday morning that the Charleston lighthouse, situated on Morris Island, and which for many years has guided the mariner to our harbor, was blown up on Wednesday night, by order of the military authorities. Nothing save a heap of ruins now marks the spot where it stood." Much of the South would be a heap of ruins by the time the war was over, and Fresnel lenses were among the casualties.

HATTERAS LIGHTHOUSE, TRAITOR TO THE STATE

The blackened coast made life difficult for the Union, which lost more ships in grounding accidents than in all of the war's naval battles. No point was more deadly than Cape Hattaras in North Carolina, which brought down nearly forty Union ships by itself. Shubrick, who had returned to head the Lighthouse Board, and Treasury Secretary Salmon Chase made it clear that recapturing the Hatteras lighthouse was a top priority. The presidents of major insurance companies had also pleaded that Hatteras was "a very important point on our coast and it is desirable the light be re-established as speedily as possible."

The Union navy thus set its sights on Cape Hatteras as the first target of the Atlantic Blockading Squadron, and sent seven gunships south in August of 1861. North Carolina had moved quickly to control the area immediately after seceding: state troops had built two forts on the southern tip of Hatteras Island near the lighthouse, which served as a lookout. Their efforts did little good. The Union fleet shelled the forts relentlessly, lobbing more than three thousand shells in the space of three hours, while the forts' ammunition fell uselessly eight hundred yards short of their targets. Both forts were taken in a matter of hours, and the Union soon

*The Twentieth Indiana Regiment camped around
the Cape Hatteras Lighthouse.*

controlled Hatteras Inlet. As an officer reported to the *New York Evening Post*, Hatteras was "a place which is of such vast importance to us. . . . With the possession of this port we can easily keep alight Hatteras lighthouse."

But the troops could not keep Hatteras alight without the lens, which they soon discovered was missing. In fact, the process of disabling North Carolina's lights had begun even before the state officially seceded. Governor John Ellis, like other Southern governors, had sent telegrams to all the principal lighthouses of North Carolina in April of 1861 directing the keepers to extinguish their lights. Most had complied, packing up the smaller lenses and sending them to the district superintendent. But taking apart a first-order lens had proved to be a daunting task for the uninitiated. After North Carolina seceded on May 20, and the Confederacy's Treasury Department officially ordered the removal of all lenses, only the one at Hatteras had remained in place. It had stayed up through the summer, prompting Confederate Major W. S. G. Andrews to complain to the governor. "Hatteras Light House is a traitor to the State, offering aid and comfort to the enemy. If it was removed," he continued, "none but those accustomed to this port could enter, and no other in these times will attempt it, and the coast would present no prominent object." Finally, at the end of July the district's lighthouse superintendent, H. F. Hancock, had sent a skilled contract machinist with a work crew to help the keeper remove the lens.

The Union missed capturing the lens by a matter of days. After taking control of the forts, the fleet's flag officer, Silas Stringham, sent Commander S. C. Rowan a message to fire up the lighthouse. "I was desirous of lighting Hatteras light," Rowan reported back, "but to my great regret I learn that the lens has been taken down and is now in Washington or Raleigh." Washington was a small town on the mainland inside Pamlico Sound, due east of Hatteras light and home to lighthouse superintendent Hancock. Commander Rowan sent an expedition there to try to retrieve the lens, but everyone the Union men spoke to claimed they did not have it, and that it had been shipped up the river "without their knowledge or consent." Rowan railed that he would demand the return of the lens before, as he put it, "I promise protection to the inhabitants" (and given the amount of pillaging going on in surrounding towns, the threat hung heavy). He insisted he would track down the guilty parties and "take all the property I can find that will reimburse the Government." And he wrote to his flag officer that with an armed boat or two, he "would go there and frighten the people out of their boots unless the lens is returned."

But the lens was no longer in Washington. Indeed it was likely that no one there knew where it was. Superintendent

Hancock had long ago fled upriver, leaving the lens unprotected in a warehouse; the warehouse owner, John Myers, was none too happy to be left holding hot property and fired off a letter to the Confederate treasury secretary, pointing out that Hancock had shirked his duty and Myers had been forced to take matters into his own hands. Only hours before the federal troops landed in Washington, he had loaded the boxes containing the lens parts onto the riverboat *Governor Morehead*, and accompanied them upriver to the town of Tarboro, where he delivered them to the army quartermaster, Captain George H. Brown. Wanting to get rid of them as soon as possible, Brown wrote to Richmond, the capital of the Confederacy, begging for an agent to be sent to take the boxes off his hands. An agent, J. B. Davidge, did show up, but his behavior prompted another letter from the quartermaster, claiming him "unfit to attend to any kind of business since his arrival. . . . I find that his habits are those of intemperance & though a delicate matter I feel my duty as an officer to furnish the department with the facts of the case." Davidge disappeared, along with the money allocated for the move. In desperation, Brown turned to Dr. David Tayloe, a physician from Washington who had been trying to join his family farther inland. If the army could furnish him a boxcar, he proposed, he would take the lens with him and hide it. On April 14, Tayloe rolled out of Tarboro, accompanied (according to the invoice) by forty-four boxes, two cases, sixty-four castings, fourteen fixtures, and two sheets of copper. After several days and considerable difficulty, he made it to Hibernia Plantation, where he rejoined his family and, he assured Martin, hid the apparatus in "a good store house in the country."

As the Union scrambled to track down the lens, the Confederacy did what it could to prevent the relighting of the lighthouse. In October, six Confederate steamships, intent on taking back Hatteras Island, landed at dawn in the tiny town of Chica-macomico on the northern end of the island. The six hundred Union soldiers stationed there, many still not fully dressed, took off in the opposite direction, running the whole length of the long, narrow sand spit, over more than twenty-three miles, on a hot day with no water. The Confederate fleet that was supposed to cut them off had, with due irony, run aground on a sand bar. The Union troops were thus able to make it to the lighthouse just after midnight, and picked up reinforcements from the Ninth New York Regiment. The two adversaries then reversed direction, with the Yankees now chasing the Confederates the length of the island back to their ships. The lighthouse remained unharmed, but the episode, named the Chicama-comico Races, prompted Treasury Secretary Chase to press the War Department for better protection for the lighthouse if they were going to entrust Hatteras with a new lens.

A steamer lamp had been hanging in the lantern room of the lighthouse since the Union troops arrived, but it was clearly not adequate for the job. Hatteras continued racking up its victims, including the celebrated USS *Monitor*, the U.S. Navy's first ironclad warship. Confederate cannonballs had bounced harmlessly off its seemingly indestructible sides during early battles, but in 1862 it passed too close to the Diamond Shoals and was swamped in its waves. Eventually, the flag officer of the Blockading Squadron announced that the lighthouse "may now be lighted with perfect safety," and the Lighthouse Board sent an engineer, W. J. Newman, to replace the lens. Surprised and delighted that the base plate, revolving mechanism, and pedestal remained, he ordered a second-order lens and had the light working by the end of May in 1862. Bateman Williams, the former assistant keeper, who had stayed on the island as a Union sympathizer, was named the new keeper. Ever optimistic in

The sinking of the ironclad USS Monitor *off of Cape Hatteras.*

these early days, the Lighthouse Board also ordered a new first-order lens from the Henry-Lepaute company, to serve as a proper replacement once the war ended.

DU PONT AND THE SOUTHERN EXPEDITION

Hatteras had been an obvious first target. But to plan the ensuing strategy, Treasury Secretary Chase requested the formation of a Blockade Strategy Board, which met during the summer of 1861, even as Union ships were retaking Hatteras Island. The Strategy Board consisted entirely of once and future Lighthouse Board members, headed by A. D. Bache, and its foremost task was to select a base of operations in the South, an important requirement given that coal-fired steamships needed frequent refueling and that it was impractical to send them north every time. The Strategy Board settled on South Carolina's Port

Royal Sound, south of Charleston, which had a good, deep entry, and further recommended dividing the Atlantic Blockading Squadron into Northern and Southern sections.

Samuel Du Pont, placed in charge of the South Atlantic Blockading Squadron, began assembling a fleet. As summer turned into fall, he scoured the North for every available ship, even borrowing the boats from Bache's Coast Survey, which were drafted into the naval ranks. With a total of seventy-seven vessels, the largest U.S. naval expedition yet assembled, the fleet left port on October 29, 1861, heading for Port Royal Sound. Four ships sank or ran aground off the South Carolina coast on the way down, but the fleet was able, after a prodigious exchange of cannon fire, to take the forts protecting the sound without losing a single ship in the battle. With a base of operations secure, Du Pont's next task was to extend the blockade and do what he could to relight the coast.

Du Pont now sent ships to the nearby town of Beaufort on Port Royal Island, the seat of lighthouse superintendent Josiah Fisher Bell, to search for lenses and other federal property. Bell had scrupulously dismantled the local lights: he spent $19.25 on blankets used to wrap the lenses, and $5.00 a month to store them in a Beaufort warehouse. He was nowhere to be found when the Union ships arrived. The town, abandoned by its white population, had been stripped of its provisions, its furniture broken, and even books and feather beds ripped apart—"a most melancholy example of weakness and dereliction." Locating some Fresnel lenses in the arsenal, Du Pont took possession of them, going out of his way to point out that these expensive pieces "belong to the United States."

His next stop was Tybee Island at the mouth of the Savannah River in Georgia. He found the island deserted and the lighthouse still standing, but its second-order lens was missing, presumably taken to Savannah. Du Pont hoisted the American flag atop the lighthouse and put up a temporary beacon. The Confederates had retreated about a mile up the river to Fort Pulaski, but in the next months a raiding party slipped out and ignited a keg of powder on the third floor of the lighthouse, rendering it useless.

Nevertheless, by the spring of 1862, the Union largely controlled the Atlantic coastline, and Du Pont set out to inventory the Southern lighthouses. The coast was uniformly dark, but behind the darkness roiled the chaos and confusion of war. Individual keepers had been responsible by and large for taking down their own lights, and their actions (like their political sympathies) ran a complicated spectrum. Many lenses were painstakingly dismounted and safely packed, either to be put on the train for safekeeping inland or to be buried in the keeper's backyard—or to be lost entirely. The revolving first-order light of Cape Romain, South Carolina, proved too troublesome to dismantle; the Union party landing there was greeted with broken prisms scattered across the ground. The lantern room had been trashed, and, as Du Pont reported, "the apparatus itself ruthlessly destroyed." Bulls Bay Lighthouse, a

The Tybee Island Lighthouse burning, after a Confederate raiding party lit a keg of powder inside.

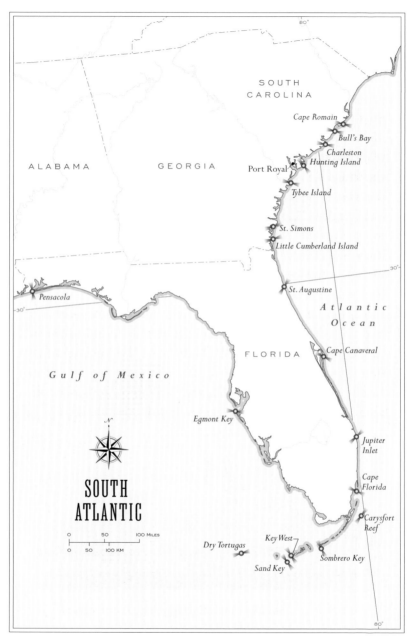

Map of the southern Atlantic coast.

few miles south, fared no better: "everything recklessly broken, down to the oil cans," read Du Pont's report. The brand new tower on Hunting Island, lit less than two years before with a second-order Fresnel lens, was rubble.

Everywhere he could, Du Pont tried to track down the missing lenses, though the keepers were generally long gone or not inclined to be helpful. James A. Clubb, however, the keeper of the light at Little Cumberland Island, Georgia, rowed out to the Union ships as soon as they entered St. Andrew Sound and described in detail how he had taken down the lens, packed it, and sent it to Brunswick on the mainland. Du Pont, accordingly, ordered a large force under cover of a gunboat to take the town. The men found Brunswick deserted, however, with no sign of the lens and no clue where it had gone.

Farther down the coast in Florida, the squadron encountered substantially less cooperation. Already the previous September, the Union navy had heard rumors of "a gang of pirates from St. Augustine" that roamed the Florida coast destroying its lights. The officer in charge of taking the town, Lieutenant J. W. A. Nicholson, wasted no time in bringing its lighthouse superintendent, Paul Arnau, in for questioning.

Jupiter Inlet Lighthouse housed a new revolving, first-order lens, which a Confederate raiding party dismantled and hid nearby.

Arnau had been very busy in the preceding weeks. He had personally seen to the dismantling of the St. Augustine lens, which he took into custody and hid. He then headed south to the next lighthouse on the coast, at Cape Canaveral, with a heavily armed group of equally passionate rebels who called themselves the "Coast Guard," rumored to be two hundred strong. After driving off the keeper there, they dismantled the lens and took it with them.

Next in line was Jupiter Inlet, with its brand-new, top-of-the-line first-order order revolving lens. On the way there, Arnau's gang had run into Augustus Oswald Lang, the lighthouse's staunchly Confederate assistant keeper who had walked off the job in disgust when the head keeper, José Francisco "Joe" Papy, refused to extinguish the light. Lang returned to Jupiter Inlet with the posse, where they informed Papy that they came "as citizens of the Confederate States to discharge a duty to our country." Papy chose not to resist the heavily armed group, who allowed him to leave in his own boat. Before he was out of sight, they began to dismantle the lens, taking some key pieces with them to render it inoperable and hiding the rest in the wilderness.

This "band of lawless persons," as the Lighthouse Board called them, continued on to the Cape Florida Lighthouse near Miami. Its keepers, however, were reported to be heavily armed and strongly pro-Union. The Confederates decided their best bet was to swoop in, destroy the lens, and get out. As one member put it, "The Light being within the immediate protection of Key West and almost indispensable at this time to the enemy's fleet, as well as knowing it to be useless for us to try and hold it, we determined to damage it so that it will be of no possible use to our enemies." They tricked the keepers into coming down from the tower by having Lang, whom the keepers knew, call up to them. As soon as the keepers opened the door, the raiders took them prisoner. The Confederates then smashed all of the lens glass and dismantled the lamps and burners, which they took with them.

Arnau's exploits, ranging over three hundred miles, had left every lighthouse on Florida's Atlantic coast inoperable. He feigned ignorance, however, at Lieutenant Nicholson's questioning. Only when Nicholson let him know that he would remain under arrest on the Union ship until he was more forthcoming did Arnau send carts into the countryside to retrieve the St. Augustine and Cape Canaveral lenses.

The Florida Keys had the only Southern lighthouses that remained lit throughout the entire conflict. Commander Louis M. Goldsborough, sent by Du Pont to check out the situation, reported that the lighthouses on Carysfort Reef, Sombrero Key, Key West, Sand Key, and Dry Tortugas were "shining as the safeguards and symbols of fraternal commerce and peaceful civilization." But the continued operation of the lights owed at least as much to the vagaries of political sentiment as to benevolence. The Keys, like much of Florida, were divided in their sympathies, and Charles Howe, the lighthouse superintendent on Key West, tilted toward the Union. It was not a popular position; in fact, when Florida seceded from the Union, he and District Judge William Marvin were the only two civil officials who did not immediately resign their federal posts. The Southern-leaning majority did not make life easy for them. But the mainland was far away, and Union ships were a constant presence. Moreover, U.S. Army Captain John Brannan had, acting on his own accord, marched his men in to occupy Key West's not-yet-completed Fort Taylor in 1861, and Key West became an important base for the blockade.

Although the Union flag flew at the Tampa Bay Lighthouse, located on an island at the mouth of the bay, the keeper's loyalties were with the South, and he smuggled the lens to Confederate forces on the mainland.

The lighthouse keepers themselves were not always happy to keep the lights shining for the Union fleet. At Key West, keeper Barbara Mabrity made it clear that her loyalty lay with the Confederacy. The Union was reluctant to fire a widow with six children who had been tending the light for almost three decades, since her husband's death. But her intransigence got her the boot in 1863 at the age of eighty-two. George Rickard, the keeper at Egmont Key, played a more complicated double game. He was stuck on the small island at the mouth of Tampa Bay, halfway up Florida's west coast. Surrounded by Union ships, he put on a friendly face, but at his first chance he dismantled the lens and smuggled it past the blockade into Tampa, making the west coast of Florida as dark as the eastern side.

THE GULF COAST LIGHTS

Jefferson Davis certainly appreciated the value of a good lighthouse. Indeed, the state of Mississippi, which he served as congressman and senator before the Confederacy elected him president, owed nearly all its coastal lights to his efforts. Like nearly all the American advocates of the Fresnel lens, Davis was a West Point graduate. In an engineering course his freshman year, he was greatly impressed with his teaching assistant, A. D. Bache, a senior about to graduate at the head of his class. Years later, Davis recalled, "He had a power of demonstration beyond that of any man I ever heard."

The two became close friends in 1845, when both found themselves in Washington, D.C., while Davis served in the House of Representatives and Bache was the newly appointed head of the Coast Survey. At one intimate party at the Coast Survey headquarters, the normally quiet and serious Davis drank enough of Bache's Rhine wine to regale the group with Indian songs he had learned out West. His social circle was exceptionally science-minded. It included Lighthouse Board members Joseph Henry and Joseph Totten, as well as Treasury secretary Robert Walker, who headed the Board. Davis looked to bring the benefits of modern technology to Mississippi, which had a single, unimpressive lighthouse at Round Island. He campaigned on the promise to better light the state's coast, and in 1847 convinced Congress to appropriate funds for three lighthouses—at Biloxi, Ship Island, and Chandeleur (today part of Louisiana).

The lighthouse at Biloxi went up quickly, its original nine brass reflectors replaced with a Fresnel lens by the start of the Civil War. Mary Reynolds, a widow raising several orphan children, was the keeper. She dutifully extinguished the light when war broke out, but the lighthouse remained a source of tension. An armed posse, including the town's mayor, calling themselves "Home Guards," demanded she relinquish the key to them, and, she suspected, they began stealing its oil for private use. Petitioning the governor for protection (and stressing that she was "a woman entirely unprotected"), she managed to stay on as keeper through the end of the war.

Farther south, on Ship Island, the stakes were higher. The site had been Jefferson Davis's pet project in Congress: he had promised to develop a navy yard there, and leaned heavily on Bache's Coast Survey as proof of its suitability. It was a great harbor, both deep and protected. Its lighthouse was lit in 1853 and outfitted with a lens in 1856; the accompanying military fort was still under construction when the war broke out. Confederate forces moved quickly to secure the fort and shore up its unfinished walls with timbers and sand bags. When the USS *Massachusetts* arrived on July 9, 1861, as part of the blockade, the light was already out. Ship and fort briefly exchanged cannon fire, and then Union Commander Melancton Smith (a lighthouse inspector before the war) ordered the ship south, to secure the lens in the Chandeleur Lighthouse before it, too, fell into rebel hands.

By September the Confederates had left Ship Island, taking the lens with them. Their parting action was to set the tower on fire. Even with a charred tower, though, this small island,

Ship Island, under Union control, as the USS Mississippi *(in the foreground) fires on a Confederate ship.*

with its deep harbor and accessible drinking water, was a strategic goldmine. Union troops moved in immediately and set about restoring the lighthouse, using an old mast from a Confederate ship to build the lighthouse's new staircase and installing a fourth-order lens salvaged from a warehouse on Lake Pontchartrain. By November 14, 1862, they had relit the light, blacking out the north-facing windows so the Confederate runners could not use it.

Ship Island was about halfway between New Orleans and Mobile, two ports that took on new importance as the Union's blockading strategy evolved. A new plan, nicknamed the Anaconda because it proposed to squeeze the Confederate forces from all sides, called for the navy to control both the Atlantic coast and the Mississippi River and work its way to the middle. New Orleans, at the mouth of the Mississippi, thus became a crucial target. And Ship Island was the natural base from which to attack it.

New Orleans was by far the largest and wealthiest city in the South, and a major American port. More than a dozen lighthouses illuminated its waters, to guide ships through the maze of passes into Lake Pontchartrain. As Union forces arrived for the blockade in the summer of 1861, they anchored ships right by the lighthouses marking the three major entries: South Pass, Southwest Pass, and Pass a l'Outre—lonely outposts in a strange landscape where marsh melted imperceptibly into ocean. The keepers were isolated and therefore unaware of the situation. Which is why, well after the Confederate call in April to extinguish the lights on the Southern coast, they had kept theirs burning. Their sympathies were not in question. Manuel Moreno, the Southwest Pass Lighthouse keeper, declared himself as "in favor of Southern rights as anyone, and shall continue so to be." But he was left without instructions:

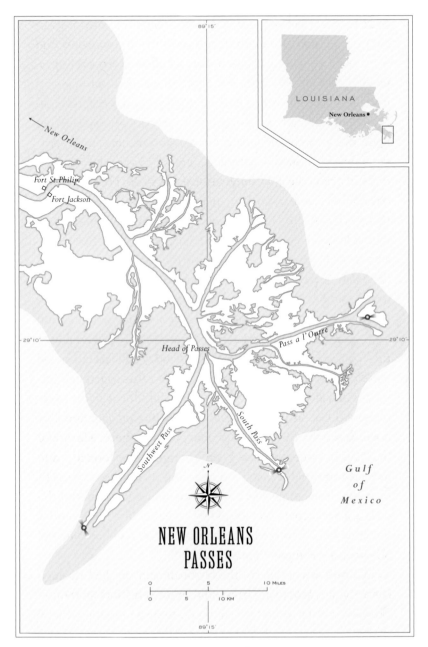

Map of the Head of Passes.

"I am in this deserted place, ignorant of what is transpiring out of it: occasionally I get a chance of procuring a newspaper, and from what a little I can learn through them, I have acquired evidence enough to induce me to believe, that we ought to have about six muskets & few pistols, and Powder & Balls, so as to be ready at all times to resist any attack that may be made upon this establishment." Left to his own devices, he improvised, blacking out one side of the lantern toward the blockaders and letting the light shine toward the shore. The man never did get his "Powder & Balls." There was little he could do when the captain of the USS *Powhatan* came ashore, accompanied by twelve men. Their visit could hardly be described as "an attack." The captain claimed they had come "for amusement" and expressed surprise that anyone was there. Their only activity was to hunt for oysters, which Moreno, perhaps with some pleasure, noted was done in the wrong spot.

Moreno no doubt hoped his half-blocked light would help the blockade runners, which typically waited in an area called Head of Passes, where the various routes came together, until they found a way to slip past the Union steamers stationed at the entrances. In June of 1861, the CSS *Sumter*, captained by none other than Raphael Semmes, paused there. He noted that the lighthouses "had been strangely overlooked," and immediately sent a boat to disable the South Pass light. His men took everything they could, including the lens and the oil. Pass a l'Outre was next. Here, they were directly under the Union guns, so the best they could do was break open the oil casks and remove the lamps.

Semmes saw a hole in the blockade and sneaked through before he was able to deactivate the Southwest Pass Lighthouse. The Union did not wait for another attempt. On July 5, thirty armed men from the *Powhatan* returned to the Southwest Pass, this time not to dig oysters. They brought a mechanic and all the necessary tools with them to dismantle the lens. "All my resistance was in vain," Moreno reported. In two to three hours the lighthouse apparatus was gone. Five days later, the USS *Brooklyn*, stationed off the Pass a l'Outre light, sent a boat to collect its lens, which Semmes had not had a chance to dismantle.

Union ships may have controlled the entrances to the passes, but the port of New Orleans, almost a hundred miles upriver, remained in Confederate hands. Forts Jackson and St. Philip, a few miles up from the Head of Passes, guarded the route to the city from opposite sides of the channel. Flag Officer David Farragut had spent months assembling a fleet on Ship Island to attack the forts, and on April 8, 1862, he lined his ships up below them and began the bombardment. After a week of steady bombing, the fleet forced its way past the forts and steamed upriver to the city itself. The lighthouses of the port and Lake Pontchartrain had largely stayed in operation until then (being too far from the coast to aid the blockaders). But as Union ships approached, Confederates removed the lenses and disabled the towers. The Bayou Bonfouca Lighthouse was so thoroughly burned that it never returned to service. In West Rigolets, the light was reactivated in November of 1862 with a ship's lantern; two days later, its keeper, Thomas Harrison, was found shot to death on the wharf outside the lighthouse. No one ever found out who did it, but the assumption was that someone was upset to see him working for the Union. He was the only lighthouse keeper to die in the line of duty during the Civil War.

The fall of New Orleans was a massive blow to the South. The only major Confederate port left in the eastern Gulf of Mexico was Mobile Bay. After securing control of the Mississippi River, Farragut turned his attention to it, relentlessly completing the slow squeeze of Anaconda. In December of 1862, Union forces took control of Sand Island, the small island in the middle of Mobile Bay that housed its major lighthouse. The light had been up only three years; previously, ships had relied on the mainland's 1822 Mobile Point Lighthouse, whose Lewis reflectors could not keep them off the shoals, which stretched some ten miles out. Sand Island Lighthouse was a beauty: built by army engineer and West Point graduate Danville Leadbetter, its two-hundred-foot tower was the tallest on the Gulf Coast and was capped with a nine-foot first-order lens.

Leadbetter eventually became a general in the Confederate army, in charge of Mobile's fortifications. In Mobile, the army concentrated its defenses at Fort Morgan, located right next to Mobile Point Lighthouse (which had been downgraded to a harbor light and outfitted with a sixth-order Fresnel lens). When the Confederates abandoned Sand Island, which they deemed indefensible, they removed the costly first-order lens as they left. The Union forces, upon taking over the island, had to content themselves with installing a new but much smaller fourth-order lens. The tower, however, was so close to Fort Morgan that the Yankees could use it to spy on activities inside the ramparts. A band of Confederates, acting on their own accord, rowed out to the island to set fire to the lighthouse. Run off by the Union ships, their leader, John W. Glenn, swore he would return and "tumble the Light House down in their teeth." And he did. On February 23, 1863, he set off seventy pounds of gunpowder underneath the tower. Little was left but scattered bricks and a pathetic stump, as Glenn proudly reported to the general in charge. We can only imagine the general's feelings: Danville Leadbetter had scarcely had time to see his most prominent engineering accomplishment in action.

*The burnt stump of Sand Island Lighthouse is visible
in the foreground. Behind it is the Mobile Point Lighthouse,
next to the smoke from the guns of Fort Morgan.*

Not all lighthouses destroyed in the Civil War were targets of such deliberate acts of Confederate sabotage. When
Farragut, now an admiral, attacked on August 5, 1864, in the
Battle of Mobile Bay, his fleet—exhorted to "Damn the tor-
pedoes!"—passed directly in front of Fort Morgan. The fort
received a good pounding, and the unfortunate lighthouse,
which found itself in the line of fire, joined its replacement
as a casualty of war.

RELIGHTING

All in all, the South darkened or destroyed 164 lights as part
of the war effort. Although the North controlled most of the
coastline by the fall of 1862, it would take years to restore the
work the Lighthouse Board had done. In the first year of the war,
only the smaller light at Hatteras and a few minor ones in
Chesapeake Bay had been relit (joining the Florida Key lights,
which had never been extinguished). Insurance companies

circulated petitions to the Lighthouse Board, requesting it to turn on the lights, and as Du Pont assured the Board that at least parts of the coast were secure enough to relight, it began to make plans. In 1863, it sent William A. Goodwin, a rather rare combination of poet and civil engineer, to survey the Atlantic and Gulf coasts and oversee the installation of new lights. His poems often featured tragic shipwrecks; his directive in 1863 was to bring them to a minimum—no easy task, as a dismaying number of lighthouses had been blown up and many others were in danger of attack. The Cape Florida light, for example, remained dark in spite of the replacement lens sitting in Key West, the workmen waiting for the moment it was safe enough to install it. Replacing Pensacola's first-order lens was "not deemed advisable until the occupancy of a greater portion of the surrounding country . . . [has] placed the station beyond risk of damage and spoliation." These fears were not unfounded: an armed party tried to demolish the relit Chesapeake lights, and at Wade Point, North Carolina, the light was reestablished only to be destroyed even more thoroughly by a "guerilla force from the mainland."

But the Lighthouse Board persevered. Slowly, the lights flickered on—nearly two dozen by 1863—facilitating the Blockading Squadron's patrol of the coastline and shining as a tangible symbol of federal power. Union control of the coast-line, so tenuous at first, was becoming more robust. Civil War historians continue to debate just how instrumental the Union blockade was in bringing down the Confederacy. Some estimate that as many as nine out of every ten runners got through in the first year of the blockade, although by the end of war the number had dropped to one in ten. The lack of supplies eventually took its toll as the South struggled to distribute food and resupply its forces. Surrender came in the spring of 1865.

The war over, some of the hidden lenses began to show up in the strangest places. Federal troops entering the deserted State Capitol in Raleigh found dozens of unmarked wooden boxes stacked higher than their heads on the second-floor balcony of the rotunda. Inside them nestled a trove of bronze castings, lamps, and clockwork: all the lighthouse equipment that had been shipped inland from North Carolina's coast. The governor had ordered the evacuation of the building just days earlier, provoking a scramble to get all of the particularly important or incriminating documents onto a train heading west. The heavy boxes filled with Fresnel lenses were left behind.

The *Philadelphia Inquirer* reported on the scene in Raleigh: "The glass concentric reflectors of Fresnel are viewed with novel curiosity by the Western men, to whom lighthouse paraphernalia is something new." Prisms were a prime find for souvenir hunters. As army engineer Montgomery Meigs, in charge of sending the equipment north, informed Shubrick, still head of the Lighthouse Board, "I learn that some broken prisms or portion of lenses have been seen in possession of boys in the streets, but the greater part of the lens apparatus will, I think, reach Washington in good order." Meigs packed everything into thirty boxes, using papers he found strewn across the Capitol floor to wrap the items. It was only well into the process that Meigs noticed the papers were revolutionary documents bearing the signatures of Thomas Jefferson, John Knox, and others. They seem to have been suitable packing material, although once the boxes arrived in Washington, they were turned over to the Treasury secretary, who gave them a more fitting place in the archives.

The Hatteras lens turned up only months later, in August, in Henderson, North Carolina. Some of the local residents had removed choice pieces from the lens, and thought the intricate glasswork would fetch a nice ransom. Instead, it brought the full attention of the U.S. Army, which quickly claimed the federal property. The lens was soon on its way to the Lighthouse Board's depot on Staten Island, New York, joining the growing number of lenses that had been found submerged in creeks, buried in the sand, or hidden in store-houses (not to mention the discarded remnants of deliberate sabotage). The twisted frames and chipped, cracked, broken, or missing prisms accumulated well beyond the capacity of the depot's attendant lamp shop, and the Board decided to return them to Paris, for Sautter and Lepaute to repair the very same lenses they had supplied throughout the 1850s. Shubrick begged the French lensmakers to work quickly, in order that "the commerce of the world be benefitted thereby." The Board also turned to a new supplier, Barbier & Fenestre, founded in 1860 by two Parisians, Frédéric Barbier and Stanislas Fenestre. The company's very first lens went to the United States, replacing the damaged fourth-order lens at the Southwest Reef Lighthouse in Louisiana. But there were over a hundred lenses to repair, and not until seven years after the war's end was the southern coastline completely relit.

Jefferson Davis remained close to the lighthouses he helped get built. He retired to the estate of Beauvoir, on the Mississippi coast, with a view of the Biloxi Lighthouse from his front porch. Beauvoir and the Biloxi Lighthouse, both constructed the same year, existed side by side for decades, the former housing the Jefferson Davis Presidential Library after his death. The two parted ways when Hurricane Katrina came ashore in 2005, leaving Beauvoir swamped by the storm surge and the Biloxi Lighthouse one of the few structures still standing on the coast.

SEVEN

THE GOLDEN AGE

THE DETRITUS LITTERING American shores in the wake of the Civil War provided, for at least one party, a glimmer of opportunity. Chance Brothers of Birmingham, England, had begun marketing its own Fresnel lenses in 1856, only to find the company completely locked out of the prodigious U.S. shopping spree. Now there was another opening. "If your Birmingham friends want to let them Lights shine in America," American naval officer H. Fuller wrote in 1866 to Fred Burdus of the American magazine *Cosmopolitan*, where Chance had run an advertisement, ". . . I have no doubt I can get a Bill introduced for its adoption next winter, but of course all this costs money: and we cannot turn the Grindstone to sharpen other men's axes for nothing."

Fuller's insinuation notwithstanding, the Lighthouse Board continued to do business with French suppliers after the Civil War. But the United States was not the only place buying lenses in the second half of the nineteenth century. With the effectiveness of the Fresnel lens universally acknowledged, it was a golden age of lighthouses worldwide, and lenses were installed by the thousands wherever ships navigated. France dominated the market, but there was enough demand for Chance Brothers to flourish as well.

Imperial Rivalries

Chance Brothers had entered the field with Jacques Tabouret's lens, which was exhibited in the Crystal Palace at the 1851 Great Exhibition in London; by 1856 the company began supplying lighthouses, animated by its rivalry with France. When Trinity House looked to improve its Europa Point Lighthouse on the rock of Gibraltar, the contract first went to Sautter, who had submitted the lowest bid. But the ensuing outcry forced Trinity House to withdraw the commission and ask Chance to make the lens instead. As British Captain Frederick Arrow complained, that would be seen by "more visitors of all nationalities in a week than any other lighthouse sees in a week of years" and should be of English manufacture.

By 1862, the *Illustrated Times* proudly proclaimed that Britain had conquered an industry that had previously existed only in France. It extolled the lens's familiar intertwining of humanitarian and utilitarian impulses: "a manufacture from which emanate the useful and the beautiful as kindred and inseparable

The Coco Islands Lighthouse in Burma, under British rule, had a first-order Chance lens.

The lighthouse exhibited in the British section of the Exposition universelle in 1867 was topped with a Chance lens.

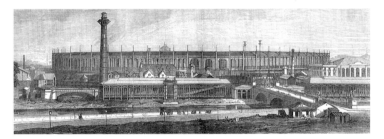

Visitors at the Pont d'Orsay entrance of the French hall at the Exposition were greeted with a Fresnel-equipped lighthouse.

spirits; where the highest faculties of the mind and the deepest sympathies of the heart have equal place; and where the genius of humanity inspires and blesses the genius of science."

Chance Brothers sent out over a thousand lenses, including more than 180 large seacoast ones. Its business remained largely within the British Empire, a worldwide market in itself. Many of the company's earliest lights went to Australia, following the discovery of gold and the development of its wool market. Others were shipped to Canada when, between 1857 and 1860, Britain footed the bill for an ambitious lighthouse-building program known as the "imperial lights," most likely inspired by the recent improvements made by its southern neighbor. Sixty Chance lenses went to India after the Crown took possession in 1858.

At the next Exposition universelle in Paris, in 1867, France and England competed head to head, each building in its national section a full-size lighthouse topped, respectively, with a Lepaute and Chance lens. Both structures towered over the other exhibits. A jury of French and British engineers compared the brightness of the two and awarded the honors to the Lepaute lens. Jules Verne, one of the seven million visitors to the Exposition, was so impressed by the design that he meticulously described a rotating Fresnel lens in his novel *The Lighthouse at the End of the World*. Verne set his tale of murder and piracy on an island off Tierra del Fuego, which indeed hosted a dioptric lighthouse by the end of the century, making it one of many remote outposts around the world to receive Fresnel's invention. From Saigon to the South Seas, from the Cape of Good Hope to the Strait of Malacca, the French and British competed to light the increasingly interconnected waterways.

TECHNOLOGICAL CHALLENGES

The two nations also competed in technology. The realization of Fresnel's catadioptric vision in the 1850s left little room to improve the optics. But the rise of steam-powered sea travel in the second half of the century posed new problems. Steamships cruised roughly twice as fast as ships under sail, which meant they needed to be warned from twice as far away. The beams were bright enough for this, but the dark period between the flashes of rotating lights was now too long. Fresnel's lights with the original formulation of eight panels flashed every minute; depending on how far away the ship was, the result could be forty seconds or more of darkness between flashes. This was adequate at the time—the first transatlantic voyage primarily under steam power didn't occur until the year of Fresnel's death—but by the 1860s, steamships were everywhere, and forty seconds was too long to leave them motoring toward the shore in the dark.

The lens manufacturers tackled the problem. The first step was to increase the number of panels to twenty-four, yielding a flash every twenty seconds. By 1868, Lepaute, by introducing rotating panels around an interior fixed lens, produced a scintillating light that flashed every four seconds. Chance Brothers, meanwhile, hired the top physics graduate from Cambridge University, John Hopkinson, to design a system known as group flashing, which divided the beam of light by setting the reflecting prisms above and below at an angle to the main beam. In every case, though, there was a trade-off. Dividing the light into several different beams allowed them to flash more frequently but inevitably weakened each beam. To compensate, lamps began to get brighter.

One response was to replace the existing lamps with an electric carbon-arc light, an innovation impressively modern but hopelessly impractical. It was in essence a miniature lightning bolt, arcing between two carbon rods hooked up to opposite ends of a battery. Physicists had known of the phenomenon since the beginning of the century, but the intense light, capable of blinding a person, had few commercial applications. When the engineers at France's central workshop

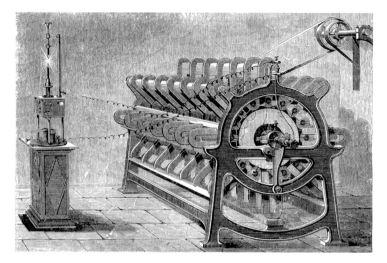

The carbon-arc lamp, shown on the left, produced a blinding light by arcing electricity between two carbon rods. A dynamo, shown on the right, was required to produce the electricity, and a steam engine (not shown but transmitting its power through the system of belts seen in the upper-right corner) was needed to power it.

experimented with carbon-arc lamps in the 1840s, they found that the lamps quickly drained the expensive batteries used to power them. (The batteries consisted of laboriously constructed piles of alternating sheets of two different metals.) The powerful generators developed in the 1850s, which created electricity through electromagnetic induction, solved that problem, but these dynamos were large devices and they needed a steam engine to power them.

England, with its big lead in steam power, tried the system out first. In 1858, Trinity House tested an arc lamp in South Foreland. Michael Faraday, who conducted the evaluation, pronounced "the fitness and sufficiency of the Magneto-Electric Light for Lighthouse purposes" and approved of its use on a permanent basis. France followed, as the Commission des phares developed a plan to use these lamps to illuminate the entire French coast. When French engineers completed the Suez Canal in 1869, they crowned the lighthouse at Port Said, the canal's northern entrance, with an electric light; as night fell during the opening celebration presided over by Empress Eugénie, a discharge of fireworks accompanied the first appearance of the lighthouse's sharp, bright beams.

The new technology proved enormously troublesome. Each arc lamp needed a generator and a steam engine, both of which required constant supervision. The boiler consumed between five hundred and two thousand pounds of fuel and four hundred gallons of water *a night*, so finding enough space to store the supplies, let alone supply the labor to feed the boiler, would be challenging. In the end, neither the French nor the British installed more than a handful of

The beams of the Port Said Lighthouse, first illuminated as part of the opening ceremony of the Suez Canal.

electric lights, placing most of them along the English Channel, as if to remind the other side that they could, if they wanted, retain these demanding lights.

Another attempt to brighten the beams came by adding more wicks to the lamp. Four wicks had seemed a lot when Augustin Fresnel and François Arago had introduced their lamp in 1820. But oil lamps gave way to kerosene and gas designs, wicks were replaced with gas jets, and the lamps grew ever larger. John Wigham, working for the Irish Lighthouse Authority, went to the extreme, designing, in 1868, a lamp with 108 gas jets arranged in concentric circles. The light it produced was intensely bright, but when he placed the lamp inside a first-order lens, the flame, more than eleven inches in diameter, was so big that much of it lay outside the focal point of the lens. This "extra-focal" light was not bent outward into the beam, and thus produced no useful effect. Wigham needed a larger lens.

Fresnel had not conceived of the need for a lens bigger than the first-order one, whose light could easily be seen until the curvature of the earth eclipsed it. In his rational plan, therefore, there was no room in the nomenclature for expansion. Wigham and Frédéric Barbier, the French lensmaker whom he approached with the idea of producing a larger lens, settled on calling it "hyper-radiant." Barbier designed a lens panel with a 1,330-millimeter (4⅓-foot) focal length, almost half again as large as a first order's 920 millimeters (3 feet). He got an opportunity to make it in 1884, at the request of the Northern Lighthouse Board's David A. Stevenson (the third generation of Stevensons there—he was the son of Alan's brother David). The Northern Lighthouse Board, along with Trinity House and the Irish Lights Authority, arranged a series of trials at South Foreland to compare the illuminating powers of electricity, gas, and kerosene. They found that electricity generated the brightest light, and oil the most economical, but the light produced by gas, Stevenson noted, was much improved with the use of the hyper-radiant panel. Present at the trials was James Kenward of Chance Brothers, who returned to Birmingham and ordered the building of an entire lens (which he renamed "hyper-radial").

The United States bought one of the very first hyper-radiant lenses from Paris in 1887. It was a fixed apparatus of 1,188

Chance Brothers's hyper-radiant lens, installed in 1909 in the Manora Point Lighthouse in Karachi, Pakistan (then part of British India). It rests on a mercury float, and its four-sided design was typical of the fast-spinning arrangement.

prisms that weighed, with its pedestal, roughly fourteen tons. Over twelve feet high, with an inside diameter of eight feet nine inches, it was the size of a room—in fact, it functioned like a room: the keepers entered inside it to tend the lens. Although originally intended for Mosquito Inlet in Florida, the lens was never installed there. Instead, the Lighthouse Board showed it at many different fairs and expositions, including the World's Columbian Exposition in Chicago in 1893, where, the Board reported, "some idea may be had of the interest taken in lighthouse building by the fact that the exhibit was one of the most popular at the fair." Truth be told, the giant lenses were almost more in demand as fair exhibits than as actual light-station beacons. When France built a hyper-radiant lens for itself in 1889, its first stop was the Exposition universelle of that year, and it continued to tour for the next five years. It took even longer for the American hyper-radiant lens to find a home, finally being dragged out of storage for Hawaii's Makapu'u Lighthouse in 1909. Although Chance and Barbier eventually made over two dozen of these giant lights between them, demand remained low. The United States never ordered another one.

Thus two significant attempts to improve upon Fresnel's design—the carbon-arc lamp and the hyper-radiant lens—were essentially expensive novelties that never fully caught on. The one innovation that did endure had already been thought of by Fresnel when he was still grumbling over Bernard-Henri Wagner's lurching escapement for rotating lenses: that floating the entire device on top of a bath of mercury would allow it to turn smoothly. "This project will not present any difficulties," he wrote, with misplaced confidence. The principle of flotation dictates that a floating body will displace a volume of liquid equal to its weight. Since first-order lenses were heavy, reaching weights of up to two tons, each bath would need two tons of mercury. When Fresnel began using the smoother machinery of Augustin Henry, he abandoned the idea of floating lenses.

The idea stayed dormant until 1890, when Léon Bourdelles, chief engineer for the Commission des phares, came up with an elegant variation. He placed the mercury inside a deep drum, with another, slightly smaller hollow drum floating on it. The heavy lens pushed down on the interior drum, nestling it inside the larger one with a layer of mercury in between. The large volumes being displaced were thus mostly inside the hollow drum, not taken up by the mercury itself. In this way, even a three-ton apparatus (not unheard of at the time) could float on about two hundred pounds of mercury.

The effect of the mercury float was everything Fresnel had dreamed of. The system was now essentially frictionless, and the slightest touch could turn the largest lenses. There was nothing stopping the lighthouse engineers from spinning lenses as fast as they wanted. With the limits of human perception as his only constraint, Bourdelles turned to the work of physiologists who claimed that the human eye only needed one-tenth of a second to perceive the full effect of a light. He designed lights with a one-tenth of a second flash that made a complete revolution in less than a minute (previously lights made one revolution every four to eight minutes). Now that the length of time between flashes was no longer an issue, Bourdelles reduced the number of panels to as few as four, thus increasing the strength of the beam. Called the *feu éclair*, this "lightning-flash light" was both as bright and as brief as its name implied. In fact, the flashes from Bourdelles's first attempts were a little too short. A tenth of a second worked fine in experiments, but on the open sea, sailors were unnerved by a light they could miss in the blink of an eye. Eventually, most *feux éclairs* settled on a flash of about three-tenths of a second.

Today the idea of an open mercury bath in every light-house would raise eyebrows, but there was little awareness of the element's toxicity at the time. Indeed, mercury was still largely associated with curative effects (Abraham Lincoln ingested it regularly). As we now know, chronic exposure to mercury vapor affects the nervous system, with symptoms that include memory loss, confusion, anxiety, and mood swings. More than one historian has noticed that this symp-tomatology neatly coincides with the lighthouse keepers' widespread reputation as madmen, drunkards, and depres-sives. Events at Rottnest Island, off Australia, bolster the case: a lighthouse built in 1896 with a first-order lens revolving in a mercury bath saw its first three lighthouse keepers take their own lives. In the United States too, from Minot's Ledge to Point Reyes, keepers were removed in strait jackets or found drunk by the side of the road. We can only guess at the mercury levels in their blood. One thing is certain: from the intense isolation to the danger of waves crashing overhead to the incessant fog whistles, there were more than enough attributes of the job to drive a person insane.

LIGHTHOUSES AT WAR

The French presented the Fresnel lens as a humanitarian gift to civilization, but by the end of its golden century its military implications had become clear. The Crimean and Opium Wars and the American Civil War had shown the stra-tegic importance of a well-lit coast. Colonial powers often installed a lens as one of their first acts of imperial acquisition. France had Fresnel lenses running in Algeria, Vietnam, and Senegal before the fighting for these areas was over. Britain followed suit, even controlling lighthouse construction for nominally independent countries, like China. Meanwhile, America's early Fresnel supporter, Matthew C. Perry, noted the best spots for future lighthouses as he first steamed forcibly into the harbors of Japan in 1853.

Other countries saw the manufacture of the lenses as a potential issue of national security, and many wanted to control the process themselves, particularly those with ambitions for the world stage. But lens manufacture, which combined precision craft glassmaking with industrial capacity, was a tough business to break into. The United States looked into it in the nineteenth century, but realized it could not produce sufficiently high-quality glass. The Japanese sent a delegation to the Chance factory in Birmingham in 1871 with the intention of learning its techniques, but abandoned the idea when they saw what was involved. By the 1880s, production remained in the hands of four companies: Chance; Henry-Lepaute; Sautter (which became Sautter, Lemonnier & Cie when Sautter partnered with civil engineer Paul Lemonnier in 1870, then Sautter Harlé when Lemonnier retired and another engineer, Henri Harlé, bought into the company in 1890); and Barbier & Fenestre (which became Barbier & Bénard in 1890, when one of Barbier's four daughters married Joseph Bénard, and then Barbier, Bénard & Turenne when Paul Turenne married another daughter in 1897).

The only country with an industrial base and glassmaking tradition to rival those of France and England was Germany. If anything, Germany's reputation for quality glass was older, and the Germans continued to produce some of the world's best lenses for microscopes and cameras. They had not attempted lighthouse lenses, most likely because the southern regions specializing in glass were politically distinct from those along the coast. Unification of the country in 1870 produced a world power eager to show its might on the world stage, and soon the optician Wilhelm Weule of Goslar began to fabricate lenses. His first products outfitted the spoils gained by Germany at the Berlin Conference of 1884, when the European powers gathered together in Germany's capital to carve up the map of Africa. Weule went on to install dozens of lenses, including first-order ones, primarily in the North Sea and among German possessions such as Qingdao and East Africa.

By the twentieth century Americans had made a name for themselves in the glass industry, specializing in the use of pressed glass and machine production of inexpensive, durable glass tableware. The U.S. government took up the issue of producing Fresnel lenses again and approached George A. Macbeth, president of Macbeth-Evans Glass, located in Pittsburgh and the leading glass manufacturer in the country, to do the work. Macbeth originally was not interested, finding the risk too high and the outcome too uncertain. But the challenge proved irresistible. He turned to the chair of the physics department at the Carnegie Institute of Technology, Harry S. Hower, to provide designs for the lenses. The Lighthouse Board installed their first lens in 1911, but their grand coup came in 1913, when they won the contract to supply the lights at both entrances of the Panama Canal, "making Pittsburg light the way from the Atlantic to the Pacific," as Macbeth put it. Macbeth-Evans stuck to making fourth-order lenses and smaller ones, but these were some of the first to be shaped completely by machine at a considerable savings. By 1914, when war shut down the European suppliers, the Lighthouse Service (which replaced the Lighthouse Board in 1910) was able to order all of its lenses from the American company.

World War I made the militarization of lighthouses explicit: the U.S. Navy took control of the Lighthouse Service. In

America, as in Europe, lightships and towers were called into the war effort for mine-laying or lookout. Despite some losses (a lightship was bombed by a German submarine in 1918), American lights generally stayed on through the First World War, being occasionally extinguished only to deter the enemy.

Lighthouses played a more dramatic role in the Second World War. The day after the bombing of Pearl Harbor, the entire American coastline went dark (although a few, such as those on Alaska's Inside Passage, were later turned back on). Not suspecting an attack over such vast distances, Canada's Pacific lighthouses stayed lit. On June 20, 1942, however, a Japanese I-class submarine surfaced off the western shore of Vancouver Island and fired over twenty shells at Estevan Point Lighthouse, with its first-order seacoast lens. This was the first attack on Canadian soil since 1812, and it was the only one on western Canada in World War II. The outcome was more comic than tragic. The lighthouse was undamaged, and the area's only inhabitants, the Nootka Indians at Hesquiat village, chased the submarine off in their canoes. But the action resonated: Canada shut down the light, which drastically curtailed shipping in the area.

The lighthouses of France suffered the most. Upon occupying France in 1940, the Germans declared the coast a military zone, off-limits to French civilians without special permission. Keepers locked out of their lighthouses reacted with horror as "the elements of occupation" marched in and out with impunity, handling the delicate optics unsupervised. On the Île Tristan, a group of drunken German sailors tried to see how fast they could make the lens spin around, with unfortunate results to the machinery. The French inspector general of lighthouses protested to the head of the German Kriegsmarine (German navy) stationed in Paris, and a certain number of keepers were allowed back to their posts, although they were never permitted to be alone in the lighthouses and had to share their already tight quarters with a representative of the Hafen Kommandanten (harbor masters).

Using French ports to make forays into the Atlantic, the Germans had an interest in keeping the lights working, although generally they kept them dark unless a German ship was arriving or departing. They also did not want the lighthouses to fall back into the hands of the Allies. In 1943, German soldiers secretly began drilling holes in the towers and filling them with dynamite, then wired the lighthouses so they could set off explosions at the first sign of attack. As the Allied forces retook the coast in the summer of 1944, the Germans generally destroyed the lighthouses before they retreated. Not even lighthouses far from the military action were spared. "It was purely due to Schadenfreude," complained the chief engineer at Cap Ferret, when he recounted how the Germans retreated without incident on August 21, 1944, then sneaked back the next night to light the dynamite.

Some lighthouses survived: Engineer T. P. E. Le Net reported this conversation with two German sailors who were guarding the lighthouse at Port-Navalo in Brittany on July 15; the Germans were pulling out, and the tower was mined and ready for dynamiting.

> [Le Net]: You are not going to blow up the light as you are leaving, are you?
> [German sailor]: We have orders to.
> [Le Net]: Lighthouses serve sailors of every country to save their lives and those of their passengers, you will not destroy the lighthouse.
> [German sailor]: Maybe.

The guards did not blow up the lighthouse, but the Germans destroyed more than a hundred towers as they retreated across France. Efforts to save the lenses had yielded few results. After taking stock of the war's damages, lighthouse authorities reported that 165 optics had to be replaced, and 25 repaired.

FRESNEL'S LEGACY

France rebuilt its lighthouses quickly, as did every other nation that suffered losses in the war. Aids to navigation, they argued, were a matter of life and death, and restoring them could not be delayed. The broken or missing Fresnel lenses were replaced, despite the cost. But the golden age of lighthouses was over. Both sides in the war had invested heavily in the new technology of radio navigation. The British and Germans each developed their own systems to guide their planes during air raids at night. The United States threw its own substantial resources behind radar, eventually developing a system that was soon expanded to civilian use after the war. Radio beacons had been supplementing visual signals in lighthouses since the 1910s. After the war, however, global systems that allowed a ship or a plane, using signals from multiple beacons, to extrapolate its location to within a few yards took over, and then this technology itself was superseded in the 1990s, as satellite-based global positioning systems became even more comprehensive and accurate.

Manufacturers ceased production of first-order seacoast lights in the years after the war. But the stepped prisms of Fresnel's original design remain downright ubiquitous, spurred by the increasingly inexpensive techniques of molding glass and plastic. Fresnel stage lights have become a staple of theaters everywhere. Stoplights, car headlights, and overhead projectors all employ the genius of his optical insight, the principle behind devices as simple as handheld sheet magnifying glasses and as new as concentrated solar generators.

But perhaps the most visible role of the Fresnel lens in the twenty-first century remains its status as an icon. Even as the U.S. decommissions many of its lighthouses, the lenses become museum centerpieces, a favorite of viewers just as they had been in the days of the universal exhibitions. They seem to evoke a bygone world of danger and romance, when getting on a ship meant complete isolation and the light from a Fresnel lens was the first sign of land. Part of the romance of the great seacoast lights was their location at the very edge of the civilized world, rising out of a landscape otherwise untouched by human presence.

Though lighthouses strike us as symbols of seclusion and remoteness, the truth is that they stand for something entirely different. No better marker of connectedness exists in this era of expanding international networks. The moment a location became important in the new global commerce was when it received a Fresnel lens. The beam of the lighthouse often signaled an end to isolation, and a link to the rest of the world. Fresnel lenses lit the way for the period known as "modern globalization," marked by an explosion in sea travel and worldwide trade.

Fresnel's lens united the major themes of burgeoning modernity: science, industrialization, national ambition. There is a well-known phrase in French that touches on the particular mixture of glory, nationalism, and global ambitions: *Faire rayonner la France*, or "make France radiant." This is precisely what Fresnel lenses did, in the most literal of ways. Making their way into the remotest corners of the world, these products of France not only shed light on the seas, but also illuminated the French system of valuing pure science and providing state support for industry. Among the most difficult pieces of technology developed in the nineteenth century, pushing at the edges of what was physically possible, the lenses required the full apparatus of this system.

At the center was Fresnel, an unlikely hero for the modern age. He worried that his position as a civil engineer lacked the grandeur of his older brother's military career, but it was civil engineers who transformed the landscape of the nineteenth century. Fresnel was part of a first generation of French engineers rigorously trained in science and deployed in the service of building a nation. Britain and the United States also saw the rise of technically proficient professional engineering corps in the nineteenth century, and it was these engineers who pushed for the adoption of his lenses.

Lighthouse work began for Fresnel as a way to get to Paris to continue his own research on theoretical physics. But the shy engineer, notoriously awkward with language, found a fluent expression in the lenses that bore his name. The most brilliant physicist of his age, he abandoned his research to devote himself to the cause of illuminating the coast. Although he was small, delicate, and hampered by sickness, the strengths be possessed were his mind and spirit, and these were enough to take on one of the civilization's most indomitable opponents: the sea. The brighter lights emanating from his lenses were one of the few human efforts to reduce the terrible risks facing every sea voyage. It is only fitting, then, that as every lighthouse in France came to have a Fresnel lens in the nineteenth century, so too did they all have a bust of Fresnel, his high, pale brow surveying the shoreline he had made a little safer.

NOTES

ABBREVIATIONS USED

AN. Archives Nationales, Paris

AS. Archives de l'Académie des sciences, Paris

CRAS. *Comptes rendus de l'Académie des Sciences*

NAB. National Archives Building, Washington, D.C.

Oeuvres. Henri de Sénarmont, Émile Verdet, and Léonor Fresnel, eds., *Oeuvres complètes d'Augustin Fresnel* (Paris: Imprimerie Impériale, 1866–1870), 3 vols.

Report. *Report of the Officers Constituting the Light-House Board* (Washington, D.C.: A. Boyd Hamilton, 1852)

RG. Record Group

INTRODUCTION: DARK AND DEADLY SHORES

11 *Méduse*: Accounts of the ill-fated Méduse include Jonathan Miles, *The Wreck of the Medusa: The Most Famous Sea Disaster of the Nineteenth Century* (New York: Atlantic Monthly Press, 2007); Alexander McKee, *Wreck of the Medusa: Mutiny, Murder, and Survival on the High Seas* (New York: Skyhorse, 2007); Albert Alhadeff, *The Raft of the Medusa* (Munich: Prestel, 2002). For a broader discussion of early nineteenth-century shipwrecks, see Jean Louis Deperthes, *Histoire des naufrages* (Paris: Ledoux et Tenré, 1815); Cyrus Redding, *A History of Shipwrecks and Disasters* *at Sea* (London: Whittaker, Treacher, 1833); John Graham, *Ship-wrecks and Disasters at Sea* (London: Routledge, 1866); Terence Grocott, *Shipwrecks of the Revolutionary and Napoleanic Era* (London: Caxton Editions, 2002); E. Chevalier, *Histoire de la Marine Française de 1815–1870* (Paris: Hachette, 1900).

11 "We abandon them!" Alexandre Corréard and J. B. Henry Savigny, *Naufrage de la fregate la Méduse* (Paris: Hocquet, 1817), 55.

14 Lloyd's of London reported: *Journal of the House of Commons* 91 (1836), 811.

14 In France, two naturalists counted: Jean Audouin and Henri Milne Edwards, *Recherches pour servir à l'hisoire naturelle du littoral de la France* (Paris: Crochard, 1832).

16 one of the most ancient technologies: Some general overviews of lighthouse history are Ray Jones, *The Lighthouse Encyclopedia* (Guilford, Conn.: Globe Pequot Press, 2004); Samuel Willard Compton and Michael J. Rhein, *The Ultimate Book of Lighthouses* (San Diego: Thunder Bay Press, 2000).

18 "The painter has assembled": Quoted in McKee, *Wreck of the Medusa*, 247.

CHAPTER ONE: DREAMS OF GLORY

21 Augustin Fresnel: For biographical information, see Charles Fabry, "La vie et l'oeuvre scientifique d'Augustin Fresnel," in *Oeuvres choisies* (Paris: Gauthier-Villars, 1938), 633–53; François Arago, "Augustin Fresnel," in *Oeuvres complètes de François Arago*, t. 1 (Paris: Gide et J. Baudry, 1855); "Centenaire d'Augustin Fresnel," *Revue d'Optique théorique et instrumentale*, no. 6 (1927), 493–570. More recent is Claude Brezinski, *Ampère, Arago, et Fresnel: trois hommes, trois savants, trois amis, 1775–1853* (Paris: Hermann, 2008). Fresnel's collected works, compiled largely under the care of his brother Léonor, are exceptionally well annotated and include an overview of his life and work by Émile

Verdet: Henri de Sénarmont, Émile Verdet, and Léonor Fresnel, eds., *Oeuvres complètes d'Augustin Fresnel* [hereafter *Oeuvres*] (Paris: Imprimerie Impériale, 1866–1870), 3 vols.

24 Jansenism: William Doyle, *Jansenism* (New York: St. Martin's Press, 2000); Dale Van Kley, *The Religious Origins of the French Revolution: From Calvin to the Civil Constitution, 1560–1791* (New Haven: Yale University Press, 1996).

25 "His memory seemed": Fabry, "La vie," 640.

26 "young man of the greatest": registre de l'École centrale de Caen, An XI, Archives départementales du Calvados, Caen, France, reproduced in François Dutour, Nicole Poirier, and Rémi Poirier, *Augustin Fresnel, 1788–1827* (Caen: Conseil Général du Calvados, 2000), 17.

26 École polytechnique: Bruno Belhoste, Amy Dahan, and Antoine Picon, *La formation polytechnicienne* (Paris: Dunod, 1994); Bruno Belhoste, Amy Dahan Dalmedico, Dominique Pestre, and Antoine Picon, eds., *La France des X: Deux siècles d'histoire.* (Paris: Economica, 1995); Janis Langins, *La République avait besoin de savants: les débuts de l'École polytechnique—l'Ecole centrale des travaux publics et les cours révolutionnaires de l'an III* (Paris: Berlin, 1987).

26 five feet three inches tall: Fiche d'immatriculation de Augustin Fresnel, 1804, Archives de l'École polytechnique, Palaiseau, France.

27 "He does not appear": comte de Montalivet, November 18, 1808, Archives Nationales, Paris [hereafter AN] f/14/2228/2, n. 136.

28 "I find nothing more tiresome": Augustin Fresnel to Léonor Mérimée, December 29, 1816, in *Oeuvres*, 1:xxviii.

28 "tormented by the worries": Augustin Fresnel to François Arago, December 14, 1816, in *Oeuvres*, 1:xxviii.

29 diffraction: For Fresnel's optical work, see Jed Buchwald, *The Rise of the Wave Theory of Light* (Chicago: University of Chicago Press, 1989); Jean Rosmorduc, Vinca Rosmorduc, and Françoise Dutour, *Les révolutions de l'optique et l'oeuvre de Fresnel* (Paris:

Vuibert, 2004); Kenneth Schaffner, *Nineteenth-Century Aether Theories* (Oxford: Pergamon Press, 1972).

32 François Arago: Maurice Daumas, *Arago. La jeunesse de la Science* (Paris: Belin, 1987); James Lequeux, *François Arago* (Paris: EDP-Sciences, 2008); François Sarda, *Les Arago: François et les autres* (Paris: Tallandier, 2002).

34 "not a single": Charles Coulston Gillispie, *Pierre-Simon Laplace, 1749–1827: A Life in Exact Science* (Princeton: Princeton University Press, 2000), 67. See also Roger Hahn, *Pierre Simon Laplace, 1749–1827: A Determined Scientist* (Cambridge: Harvard University Press, 2005). For a general discussion of scientific culture in the revolutionary and Napoleonic eras, see Charles Coulston Gillispie, *Science and Polity at the End of the Old Regime* (Princeton: Princeton University Press, 1980); Maurice Crosland, *The Society of Arcueil: A View of French Science at the Time of Napoleon I* (Cambridge: Harvard University Press, 1967); Lorraine Daston, "Nationalism and Scientific Neutrality under Napoleon," in Tore Frängsmyr, ed., *Solomon's House Revisited: The Organization and Institutionalization of Science* (Canton, Mass.: Science History Publications, 1990), 95–119; Jean Dhombres and Nicole Dhombres, *Naissance d'un pouvoir: Sciences et savants en France (1793–1824)* (Paris: Payot, 1989); Robert Fox, *The Culture of Science in France, 1700–1900* (Aldershot: Variorum, 1992); Ken Alder, *Engineering the Revolution: Arms and Enlightenment in France, 1763–1815* (Princeton: Princeton University Press, 1997).

34 "in short carried the spirit": *Correspondance de Napoléon ler*, 30 (1870), 330.

35 "Sire, I have no need": The words are probably apocryphal, but the spirit of the exchange was recorded in William Herschel, *The Scientific Papers of Sir William Herschel* (London: Royal Society and Royal Astronomical Society, 1912), 1:1xii.

35 "but what was my disenchantment": François Arago, *Histoire de ma jeunesse* (Brussels: Kiessling, Schnée, 1854), 55.

36 "was active and incessant": Ibid., 146.

36 "tall commanding form": *Penny Cyclopaedia* (London: Knight, 1858), 33.

37 "*ardeurs méridionales*": François Arago to Mme Pelletier, Dossier Arago, Archives de l'Académie des Sciences, Paris [hereafter AS].

38 "show[ing] himself constantly": *Extrait de l'arrêté pris à Valence le 9 mai 1815 par M. le Baron Bourdon Vatry*, AN f/14/2228/2, n. 56.

38 "When one knows Mr. Fresnel": Pierre Charles Le Sage to comte de Molé, May 30, 1815, AN f/14/2228/2, n. 53.

39 "the residence in which": Pierre Charles le Sage to comte de Molé, July 24, 1815, AN f/14/2228/2, n. 50.

39 "the poor state of my health": Augustin Fresnel to comte de Molé, July 27, 1815, AN f/14/2228/2, n. 47.

41 "leaving no doubt": Augustin Fresnel, in *Oeuvres,* 1:20

41 "My health has been weakened": Augustin Fresnel to comte de Molé, September 15, 1815, AN f/14/2228/2, n. 20.

42 "very remarkable for their novelty": François Arago to Gaspard de Prony, December 19, 1815, AS.

42 "will be profitable": Gaspard de Prony to comte de Molé, December 20, 1815 AN f/14/2228/2, n. 60.

42 "very remarkable experiments": comte de Molé to Augustin Fresnel, October 27, 1815, AN f/14/2228/2.

42 "attached the greatest importance": Augustin Fresnel to Léonor Fresnel, February 18, 1816, Document LIX/15, in *Oeuvres,* 2:833.

42 "You could not give": comte de Molé to Augustin Fresnel, June 7, 1816, AN f/14/2228/2, n. 31.

43 "I have decided to remain": Augustin Fresnel to Léonor Fresnel, September 25, 1816, Document LIX/18, in *Oeuvres,* 2:837.

43 "all pale with hunger": Émile Souvestre, *Les Derniers Bretons* (Paris: Charpentier, 1836), 190.

43 "I am a third of the time": Augustin Fresnel to comte de Molé, April 24, 1817, AN f/14/2228/2, n. 34.

43 "I burn to leave Rennes": Augustin Fresnel to François Arago, February 9, 1817, Dossier Fresnel, AS.

44 "impatient": Augustin Fresnel to comte de Molé, April 24, 1817, AN f/14/2228/2, n. 34.

44 "on every occasion treated Fresnel": Arago, "Fresnel," in *Biographies of Distinguished Scientific Men*, W. H. Smyth, trans. (Boston: Ticknor and Field, 1859), 181.

45 "the motions of the rays": Announcement in *Oeuvres*, 1:xxxvi.

45 "enemy attack": Léonor Mérimée to Augustin Fresnel, March 6, 1817, Document LIX/23, in *Oeuvres*, 2:842.

47 "just as illuminated": *Oeuvres*, 2:365.

CHAPTER TWO: THE FLASH OF BRILLIANCE

49 "on the universality of the coasts": *Oeuvres*, 3:xxii.

52 Argand: John J. Wolfe, *Brandy, Balloons, and Lamps: Aimé Argand, 1750–1803* (Carbondale: Southern Illinois University Press, 1999).

53 up to eight mirrors: Jean-Baptiste Biot and François Arago, *Recueil d'observations géodesiques astronomiques et physiques* (Paris: Courcier, 1821), 158.

53 "Sirs, I have the honor": Louis Becquey to Commission, July 21, 1819, in *Oeuvres*, 3:xxiv.

56 "the hassle of these little details": Augustin Fresnel to Léonor Fresnel, 45 2, 1819, in *Oeuvres*, 3:386.

56 "perceived at first glance": François Arago, "Fresnel," in *Biographies of Distinguished Scientific Men*, W. H. Smyth, trans. (Boston: Ticknor and Field, 1859), 269.

58 dioptric: For the invention of the lens, see Augustin Fresnel, "Phares et appareils d'éclairage," in *Oeuvres*, vol. 3, 1–474. For the technical development of the lens, see the series of articles by Thomas Tag titled "Producers of the Fresnel Lens" in *The Keeper's*

Log: "Part I," 21, no. 3 (Spring 2005); "Part II," 21, no. 4 (Summer 2005); "Part III," 22, no. 1 (Fall 2005); "Part IV," 22, no. 2 (Winter 2005); "Part V," 22, no. 3 (Spring 2006); "Part VI," 23, no. 1 (Fall 2006); also Julia Elton, "A Light to Lighten Our Darkness: Lighthouse Optics and the Later Development of Fresnel's Revolutionary Refracting Lens, 1780-1900," *International Journal for the History of Engineering and Technology* 79, no. 2 (July 2009), 183-244.

59 "broken through an open door": *Oeuvres*, 2:832.

62 François Soleil: Paolo Brenni, "Soleil, Duboscq, and Their Successors," *Bulletin of the Scientific Instrument Society*, no. 51 (1996), 7-16; Allen Simpson, "François Soleil, Andrew Ross and William Cookson: The Fresnel Lens Applied," *Bulletin of the Scientific Instrument Society*, no. 41 (1994), 16-19.

64 "dazzled by the spectacle": Augustin Fresnel to Léonor Fresnel, June 19, 1820, in *Oeuvres*, 3:387.

65 "It must have been a hardy soul": Augustin Fresnel to Léonor Fresnel, April 15, 1821, in *Oeuvres*, 3:388.

66 "enchanted": Ibid.

66 "partial and little representative": Vincent Guigueno, *Au Service des phares: La signalisation maritime en France XIXe-XXe siècle* (Rennes: Presses universitaires de Rennes, 2001), 57.

66 "not hesitate to recognize": Isaac Bordiet-Marcet to Louis Becquey, AN F14.

66 "I don't know": Augustin Fresnel, to Léonor Fresnel, April 15, 1820, in *Oeuvres*, 3:389.

71 Jacques Tabouret: Although Tabouret's first name is absent from all French documents, it appears in his contract with Chance Brothers. "An Agreement made the 1st day of January 1850," BS6/3/3/3/36, Sandwell Community History and Archives, Smethwick, England.

72 "like the slats on a Venetian blind" "Mémoire sur un Nouveau Système de Phares," in *Oeuvres*, 3:125.

72 "a good number of honest Parisians": Augustin Fresnel to Léonor Fresnel, July 25, 1822, in *Oeuvres*, 3:391.

75 six months' worth of food and supplies: Thomas Adolphus Trollope, *A Summer in Western France* (London: Henry Colburn, 1841), 2:243.

76 "Having almost only bad workers": Augustin Fresnel to Thomas Young, September 16, 1823, in *Oeuvres*, 2:764.

77 "It's only by following the experimental method": Augustin Fresnel to Robert Stevenson, June 3, 1825, in *Oeuvres*, 3:423.

CHAPTER THREE: THE DREAM OF TOTAL REFRACTION

82 Carte des phares: Augustin Fresnel and Admiral Rossel, "Rapport contenant l'exposition du système adopté par la commission des phares pour éclairer les côtes de France," in *Oeuvres*, 3:241. For the history of the French lighthouse service, see Léonce Reynaud, *Mémoire sur l'éclairage et le balisage de côtes de France* (Paris: Imprimerie imperial, 1865); Vincent Guigueno, *Au Service des phares: La signalisation maritime en France XIXe–XXe siècle* (Rennes: Presses Universitaires de Rennes, 2001); Francis Dreyer and Jean-Christophe Fichou, *L'histoire de tous les phares de France* (Rennes: Éditions Ouest-France, 2005): Jean-Christophe Fichou, Noël Le Hénaff, and Xavier Méval, *Phares: histoire du balisage et de l'éclairage des côtes de France* (Douarnenez, France: Chasse-Marée, 1999); Fichou, *Gardiens de phares: 1798–1939* (Rennes: Presses Universitaires de Rennes, 2002); Xavier Méval, *Phares de France* (Douarnenez, France: Chasse-Marée, 2006); Émile Allard, *Les phares* (Paris: Rothschild, 1889); Guy Prigent, *Phares et balises* (Rennes: Éditions Apogée, 2002); Olivier Chapuis, *A la mer comme au ciel. Beautemps-Beaupré et la naissance de l'hydrographie moderne* (Paris: Presses Universitaires de la Sorbonne, 2000).

84 Soleil for seven thousand francs: mémoire des fournitures et ouvrages faits pour le sr. Soleil (1824), AN F/14/19830.

87 found his prisms inexact: *Oeuvres*, 3:368.

87 "It is quite difficult to": Augustin Fresnel to Robert Stevenson,

1825, in Guigueno, *Au Service*, 56. From National Library of Scotland (NLS), Ace 10706 (Stevenson Papers), n. 80, Incoming letters (1824–1826).

88 "I waited some days": Augustin Fresnel to M. Georges Bontemps, director of the Choisy-le-Roi factory, August 4, 1826, reproduced in Thomas Tag, "Producers of the Fresnel Lens, Part I: The Early Development of the Fresnel Lens," Keeper's Log 21, no. 3 (Spring 2005).

88 "Monsieur Soleil complains": Augustin Fresnel to M. Tassaert, Paris, June 26, 1825, quoted in ibid.

90 "neither the leisure nor the health": Augustin Fresnel to M. Roard, Paris, December 30, 1826, in *Oeuvres*, 3:381.

91 "If one questions Arago": Augustin Fresnel to Claude-Louis Mathieu, July 18, 1822, Dossier Mathieu, AS.

91 "the most glorious and Undividable": "A Translation of the Charter of K.H. the 8th, Exemplified by His Most Excellent Majesty King George II," in *The Royal Charter of Confirmation Granted by His Most Excellent King James II to the Trinity House of Deptford-Strond* (London: Trinity House, 1763), 104.

91 built under private contract: The English example was central to the argument that lighthouses do not always function as a public good, in R. H. Coase, "The Lighthouse in Economics," *Journal of Law and Economics* 17, no. 2 (1974), 357–376; see also James Taylor, "Private Property, Public Interest, and the Role of the State in Nineteenth-Century Britain: The Case of the Lighthouses," *Historical Journal* 44, no. 3 (September 2001), 749–771. The literature, however, does not take into consideration the impact of the Fresnel lens, which ended the era of privately owned lighthouses.

91 "the evil of having lights": Quoted in Bella Bathurst, *The Lighthouse Stevensons: The Extraordinary Story of the Building of the Scottish Lighthouses by the Ancestors of Robert Louis Stevenson* (New York: HarperCollins, 1999), xxi.

92 "like a star of the first magnitude": Captain Henry Kater (second

British officer assisting Colby), "An Account of the Trigonometrical Operations in the Years 1821, 1822, and 1823, for Determining the Difference of Longitude between the Royal Observatories of Paris and Greenwich," *Philosophical Transactions* 118 (January 1828), 154.

92 Northern Lighthouse Board: Bathurst, *Lighthouse Stevensons*; Alison Morrison-Low, *Northern Lights: The Age of Scottish Lighthouses* (Edinburgh: National Museums Scotland, 2010); Alan Stevenon, *A Rudimentary Treatise on the History, Construction, and Illumination of Lighthouses* (London: John Weale, 1850); Thomas Stevenson, *Lighthouse Illumination* (London: John Weale, 1859); David Stevenson, *Lighthouses* (Edinburgh: Adam and Charles Black, 1864).

93 "Monsieur Frenel": Cited in Guigueno, *Au Service*, 29.

94 "fractious to the extent": Henry Cockburn, *Circuit Journey* (Edinburgh, 1888), 234, cited in Alison Morrison-Low, "Brewster and Scientific Instruments," in A. D. Morrison-Low and J. D. Christie, eds., '*Martyr of Science*': *Sir David Brewster, 1781–1868* (Edinburgh: Royal Scottish Museum, 1984).

94 "the Creed of the Society": David Brewster to H. Brougham, September 26, 1847, Brougham Papers, Ms. 26632, University College, London, cited in G. N. Cantor, *Optics after Newton: Theories of Light in Britain and Ireland, 1704–1840* (Manchester: Manchester University Press, 1983), 176.

94 Brewster's immediate response: Bathurst, *Lighthouse Stevensons*, 137.

94 "bizarre disposition of spirit": François Arago, "Sur le nouveau Système d'éclairage des phares adopté en France," *Annales de chimie et de physique*, t. 37 (Paris: Crochard, 1828), 399.

CHAPTER FOUR: RACE TO PERFECTION

97 "*les passions tristes*": David Barnes, *The Making of a Social Disease: Tuberculosis in Nineteenth-Century France* (Berkeley: University of California Press, 1995), 25.

98 Becquey quickly agreed to the request: Louis Becquey to Augustin Fresnel, March 12, 1827, in *Oeuvres*, 3:441 ("deplorable state" is editor's phrase).

98 "the most beautiful crown": François Arago, *Biographies of Distinguished Scientific Men*, W. H. Smyth, trans. (Boston: Ticknor and Field, 1859), 279.

99 "I could have wished to live longer": Ibid., 278.

99 "paternal kindness": Léonor Fresnel to Louis Becquey, July 25, 1827, AN f/14/2228/2, n. 163.

100 Industrial Revolution: For the history of industrialization in France, see Jean-Pierre Daviet, *La Société industrielle en France (1814–1914)* (Paris: Seuil, 1997); Antoine Picon, *L'Invention de l'ingénieur modern* (Paris: Presses de l'École nationale des ponts et chaussées, 1992); Jeff Horn, *The Path Not Taken: French Industrialization in the Age of Revolution, 1750–1830* (Cambridge: MIT Press, 2006); François Caron, *La dynamique de l'innovation: Changement technique et changement social (XIX–XXe siécle)* (Paris: Gallimard, 2010).

102 "a triumph": Cited in François Sarda, *Les Arago: François et les autres* (Paris: Tallandier, 2002), 112.

102 "Modern lighthouses": François Arago, "Les phares," in *Oeuvres complètes de François Arago* (Paris: Gide et J. Baudry, 1855), 6:56.

102 "our most useful": François Arago, June 6, 1833, in *Archives parlementaires de 1787 à 1860: recueil complet des débats* (Paris: Paul Dupont, 1892), 625.

103 "One day": "Discours d M. Arago, dans la Chambre des députés, Séance du 25 mai 1842," in *Annales maritimes et coloniales* (Ministère de la marine de des colonies, 1842), 1058.

104 "notable advantages": Quoted in Bella Bathurst, *The Lighthouse Stevensons: The Extraordinary Story of the Building of the Scottish Lighthouses by the Ancestors of Robert Louis Stevenson* (New York: HarperCollins, 1999), 153.

104 Léonor petitioned the Ponts et chaussées: Léonor Fresnel to Louis Becquey, July 4, 1837, AN f/14/2228/2, n. 85.

105 his new bride: Eulalie Françoise Réal-Lacuée was the daughter of revolutionary politician Pierre François Réal. A forty-six-year-old widow, she had received a marriage proposal from Stendhal in October of 1836, before marrying Léonor in December. Robert Alter, *A Lion for Love: A Critical Biography of Stendhal* (New York: Basic Books, 1979), 237.

105 singing parts of an opera: Léonce Reynaud to François Soleil, Paris, December 1, 1841, AN f/14/20813.

106 Léonor began pushing French manufacturers: *Commission des phares,* registre D, 5, cited in Vincent Guigueno, *Au Service des phares: La signalisation maritime en France XIXe–XXe siècle* (Rennes: Presses Universitaires de Rennes, 2001), 59.

106 Santander Lighthouse: According to his son, Lepaute first asked Soleil to provide the lenses. Soleil, however, refused, wanting "the market to be under his name alone." Lepaute, stuck with the order from Spain, was forced to try his hand at making the optics himself. Henry-Lepaute, fils to Léonce Reynaud, Paris, March 6, 1874, AN f/14/20816.

106 "a happy concurrence": Léonor Fresnel, "Note pour l'Exposition de 1844," AN f/14/20816.

109 "grave and perilous": "Note sur l'appareil catadioptrique excut par M. François jeune, pour le phare écossais de Scherivore," read January 8, 1844, François Arago, rapporteur, *Comptes rendus de l'Académie des sciences* [hereafter *CRAS*] 18 (1844), 7.

110 "determined to put in practice": Alan Stevenson, *Account of the Skerryvore Lighthouse* (Edinburgh: Adam and Charles Black, 1848), 177.

110 "heavily weight the consequences": "*Note* sur l'appareil catadioptrique."

110 Eddystone: Mike Palmer, *Eddystone: The Finger of Light* (Woodbridge: Seafarer, 2005); for English lighthouses more generally, see W. H. Davenport Adams, *Lighthouses and Lightships* (London: T. Nelson and Sons, 1870); Lynn F. Pearson, *Lighthouses* (Buckinghamshire: Shire, 2003).

110 "Went to Mr. H. Laponte": Bence Jones, *Michael Faraday: The Life and Letters of Faraday* (Philadelphia: J. B. Lippincott, 1870), 223. For more on Faraday, see Geoffrey Cantor, *Michael Faraday, Sandemanian and Scientist* (New York: Macmillan, 1991); Frank A. J. L. James, *Michael Faraday: A Very Short Introduction* (Oxford: Oxford University Press, 2010).

111 "experimented with the greatest care": Henry-Lepaute, fils to Léonce Reynaud, Paris, March 6, 1874, AN f/14/20816.

112 "sublime simplicity": Jules Michelet, *The Sea* (New York: Rudd and Carleton, 1861), 99.

116 "the installation of a 1st order apparatus": Léonce Reynaud à l'ingenieur ordinaire, June 23, 1852, AN f/14/19896, n. 45.

116 "For the sailor who steers by the stars": Michelet, *Sea*, 97.

118 took center stage: Armand Audiganne, *L'Industrie contemporaine, ses caractères et ses progrès* (Paris: Capelle, 1856).

120 "eminently national industry": *Visites et études de S.A.I le Prince Napoléon au Palais de l'industire ou Guide pratique et complèt à l'exposition universelle de 1855* (Paris: Perrotin, 1855).

120 "The invention of these devices": Ibid.

121 introduced the idea in an 1849 talk: Stevenson's presentation to the Royal Scottish Society of Arts is mentioned in its *Transactions of the Royal Scottish Society of Arts* (Edinburgh: Neill, 1851), 3:283; the first published account is Thomas Stevenson, *Holophotal System of Illuminating Lighthouses* (Edinburgh: Neill, 1851).

121 "the first lighthouse": Thomas Stevenson, *Lighthouse Construction and Illumination* (London: E. and F. N. Spon, 1881), 89.

123 "The glory of the invention": "Chronique," *Annales des ponts et chaussées* (Paris: Victor Dalmont, 1855), 3e série, 2e semestre, 86.

123 Léonor Fresnel wrote an open letter to Arago: Léonor Fresnel to François Arago, August 31, 1852, *CRAS* 35 (1852), p. 364; see also François Arago, "Sur le nouveau Système d'éclairage des phares adopté en France, Examen d'une Réclamation que le Dr. Brewster vient de faire à ce sujet" *Annales de chimie*, t. 37 (1853), 392–409.

123 Reynaud went even further: Léonce Reynaud, "Réponse de M. Reynaud à M. Stevenson," *Annales des ponts et chaussées* (Paris: Victor Dalmont, 1855), 3e série, 2e semestre, 95.

123 "As for the catadioptric mirror": Ibid., 96.

124 "You have rendered": Robert Louis Stevenson to Charles Baxter, February 2, 1873, in Ernest Mehew, ed., *Selected Letters of Robert Louis Stevenson* (Princeton: Princeton University Press, 2001), 29.

124 "whenever I smell salt water": Robert Louis Stevenson, "Memoirs of Himself," in *Memories and Portraits, Memoirs of Himself and Selections from his Notebook* (Whitefish, Mont.: Kessinger, 2003), 149.

124 "But there Eternal granite hewn": Robert Louis Stevenson, "Skerryvore," in *Underwoods* (London: Chatto and Windus, 1887), 69.

CHAPTER FIVE: THE AMERICAN EXCEPTION

127 tonnage American ships carried: In 1790, the merchant marine shipped 478,377 tons; in 1810, the tonnage was 1,424,783. No. 86, "Statement Showing the Amount of Registered, Enrolled, and Licensed Sail and Steam Tonnage," *Congressional Edition* (Washington, D.C.: Government Printing Office, 1882), 2025:297.

127 "hardly anyone in the United States": Alexis de Tocqueville, *Democracy in America*, Henry Reeve, trans. (New York: J. and H. G. Langley, 1840), 2:42.

128 West Point: R. Ernest Dupuy, *Where They Have Trod: The West Point Tradition in American Life* (New York: Frederick A. Stokes, 1940); Ann Johnson, "Material Experiments: Environment and Engineering Institutions in the Early American Republic," *Osiris* 24 (2009), 53–74.

129 U.S. lighthouse system: For a history of American lighthouses, see Arnold Burges Johnson, *The Modern Light-house Service*

(Washington, D.C.: Government Printing Office, 1890); Francis Ross Holland, *America's Lighthouses: An Illustrated History* (Brattleboro, V.: Stephen Greene Press, 1972); Dennis Noble, *Lighthouses and Keepers* (Annapolis, Md.: Naval Institute Press, 1997); Elinor De Wire, *Guardians of the Lights* (Sarasota, Fla.: Pineapple Press, 1995); Francis Ross Holland Jr., *Great American Lighthouses* (Washington, D.C.: Preservation Press, 1989); David Cipra, *Lighthouses and Lightships of the Northern Gulf of Mexico* (Washington, D.C.: Government Printing Office, 1976); James Gibbs, *Lighthouses of the Pacific* (West Chester, Penn.: Shiffer 1986); Ray Jones, Bruce Roberts, and Cheryl Shelton-Roberts, *American Lighthouses: A Comprehensive Guide to Exploring Our National Coastal Treasures* (Guilford, Conn.: Globe Pequot Press, 2012).

129 Commerce Clause: Adam Grace, "From the Lighthouses: How the First Federal Internal Improvement Projects Created Precedent That Broadened the Commerce Clause, and Affected Early Nineteenth Century Constitutional Debate," *Albany Law Review* 68 (2004), 97.

130 "unnecessary alarm": S. Pleasonton to W. H. Winder Jr., August 7, 1848, reproduced in Gaillard Hunt, "The History of the Department of State: VI," *American Journal of International Law* 4 (July 1910), 596, 611, 603.

130 "being fatigued with the ride": Ibid.

134 "tall, fine-looking man" *New England Historical and Genealogical Register for the Year 1863* (Albany: J. Munsell, 1863), 17:4.

134 "I could not fetch Province town": Winslow Lewis to Stephen Pleasonton, April 6, 1842, Letters from Winslow Lewis, Records of the Coast Guard, Record Group [hereafter RG] 26, National Archives Building [hereafter NAB], Washington, D.C.

135 "mathematical instrument maker": *New England Historical and Genealogical Register*, 17:4.

136 "a bad light worse": "Lighthouse Establishment," *Appletons' Annual Cyclopaedia and Register of Important Events* (New York: D. Appleton, 1887), 444.

138 "The money saved": H. R. Doc. No. 28-38 (1844).

139 "great deficiency in our present": E. and G. W. Blunt to Treasury Secretary Duane, September 18, 1833, RG26, E31, Lighthouse Letters series P (1833–1864), Box 1 (1833–1837), 88, NAB.

140 "nothing but the Argand lamp": Communication from Messrs. E. and G. W. Blunt, November 30, 1837, in *Senate Documents, 32nd Congress, 1st Session* (Washington, D.C.: Boyd Hamilton, 1852), 6:496.

140 "by full and satisfactory experiment": "Extract from the Report of the Board of Officers Convened under the Direction of Hon. Thomas Corwin, Secretary of the Treasury, on the Light House System of the United States Coast," *American Journal of Science and Arts*, 2nd series, 13 (May 1852), 320. Quoted in *American Journal of Science* 63, 320.

140 "You need be under no apprehension": Stephen Pleasonton to Winslow Lewis, August 12, 1840, Miscellaneous letters, RG26, 17H, Box 14, NAB.

141 "If the experiment shall be successful": John Davis to Stephen Pleasonton, March 21, 1840, Miscellaneous letters, RG26, 17F, Box 1, NAB.

141 for the same amount of oil: S. Doc. No. 25-428 (1838).

141 could be seen only ten to sixteen miles away: Extracts from S. Doc. No. 26-474, at 529 (May 18, 1840).

142 "I should be glad": John Davis to Stephen Pleasonton, Washington, Friday, 1840, Miscellaneous letters, RG26, E17F, Box 1, NAB.

142 "the best light ever seen": Ibid.

142 "Mr. Pleasanton has purposely thrown": Matthew Perry to E. Vail, June 15, 1839, Perry Papers, Library of Congress, cited in John Schroeder, *Matthew Calbraith Perry: Antebellum Sailor and Diplomat* (Annapolis, Md.: Naval Institute Press, 2001), 85.

142 "The truth is the old egotist": Matthew Perry to E. Vail, August 11, 1839, Perry Papers, Houghton Library, Harvard

University, Cambridge, Mass., cited in Schroeder, *Matthew Calbraith Perry*, 86.

142 "Having only received": Louis Bernard, chef d'atelier de la maison Henry-Lepaute, to William Hoyt, New York, August 13, 1840, Miscellaneous letters, RG26, 17F, Box 1, NAB.

143 Pleasonton wrote to Davis: Stephen Pleasonton to John Davis, Draft Copies of Letters Sent, RG26, E17H, Box 14, NAB.

143 lantern room was too small: Stephen Pleasonton to John Davis, RG26, **118**, Box 14, NAB.

143 Bernard gave a full account: Louis Bernard to John Davis, Navesink, December 11, 1840, Miscellaneous letters, RG26, E17F, Box 1, NAB.

143 "ordinary mechanics or laborers": S. Exec. Doc. No. 32-28, at 581–582 (1851–1853).

144 "The lenticular lights at the Navesink": Extracts from H. R. Doc. No. 27-183 (February 25, 1843).

144 "on the Fresnel plan": *American Coast Pilot* (1847), 208.

144 "a thorough examination by competent": New York, June 2, 1842, Extracts from H. R. Doc. No. 27-183, at 539 (February 25, 1843).

145 "Instead of order, economy and utility": Winslow Lewis to Stephen Pleasonton, April 6, 1842, Letters from Winslow Lewis, RG26, 17E, Box 2, NAB.

145 "236 witnesses": I.W.P. Lewis to Robert Winthrop, quoted in ibid.

145 "no man of science": Ibid.

146 "an idler or an obscure individual": Ibid.

146 "an inexperienced individual": Winslow Lewis to Stephen Pleasonton, April 6, 1842.

146 "his whole knowledge": Winslow Lewis to John C. Spencer, April 27, 1843, R26, E31, Box 2, #44, NAB.

146 "base and scandalous attack": Winslow Lewis to Stephen Pleasonton, April 6, 1842.

146 "malicious & in general confounded": Stephen Pleasonton to John C. Spencer, March 6, 1843, RG26, E31, Box 2, #22, NAB.

146 he had fitted only four lighthouses: Stephen Pleasonton to John C. Spencer, R26, E31, Box 2, #122, NAB.

147 "praying for the establishment": Merchants and Underwriters petition, March 26, 1849, RG26, E31, Box 5, #361, NAB.

147 "obvious and easy to be understood": "Report of Professors Pierce and Lovering, of Harvard University, on Fresnel's Dioptric Apparatus for Lighthouses," *Journal of the Franklin Institute* 48 (February 1846), 250.

147 on I.W.P.'s request: I.W.P. Lewis to M. Meredith, July 2, 1849, RG26, E31, Box 5, #351, NAB.

147 "a fatal spot": "Report of I.W.P. Lewis, Civil Engineer, Made by Order of Hon. W. Forward, Secretary of the Treasury, on the Conditions of the Lighthouses, Beacons, Buoys, and Navigation, upon the Coasts of Maine, New Hampshire, and Massachusetts, in 1842," in *Compilation of Public Documents and Extracts from Reports and Papers Relating to Light-Houses, Light-Vessels, and Illuminating Apparatus, and to Beacons, Buoys and Fog Signals, 1789–1871* (Washington, D.C.: Government Printing Office, 1871), 349.

148 "the rocket light": W. R. Easton, Customs Collector, to Thornton Jenkins, August 3, 1851, "Letters Relating to Sankaty Head," in *Executive Documents Printed by Order of the Senate of the United States during the First Session of the Thirty-Second Congress, 1851–2* (Washington, D.C.: A. Boyd Hamilton, 1852), 6:338.

149 "It was an event": Richard Meade Bache, *Life of General George Gordon Meade* (Philadelphia: H. T. Coates, 1897), 561.

149 "We advise shipmasters": *American Coast Pilot* (1847), 256.

149 ordered two first-order lenses from Paris: I.W.P. Lewis to Col. J. J. Abert, November 5, 1850, RG26, E16, Box 2, NAB.

150 custom officials followed their standard policy: William Allen Butler and Harriet Allen Butler, *A Retrospect of Forty*

Years, 1825–1865 (New York: Charles Scribner's Sons, 1911), 199–200.

151 Emerson, speaking: Ralph Waldo Emerson, "The Young American: A Lecture Read before the Mercantile Library Association, Boston, February 7, 1844," in *Nature; Addresses and Lectures* (Boston: James Munroe, 1849), 349.

151 "ample geography": Ralph Waldo Emerson, "The Poet," in *Essays: Second Series* (Boston: James Monroe, 1844), 41.

153 "inhabitants almost of an other planet": Quoted in Albert E. Moyer, *Joseph Henry: The Rise of an American Scientist* (Washington, D.C.: Smithsonian Institution Press, 1997), 214. For more on Henry and early American science, see Robert V. Bruce, *The Launching of Modern American Science, 1846–1876* (Ithaca: Cornell University Press, 1988); Sally Gregory Kohlstedt, *The Formation of the American Scientific Community: The American Association for the Advancement of Science, 1848–60* (Urbana: University of Illinois Press, 1976).

153 "The effect is very brilliant": Joseph Henry, *The Papers of Joseph Henry*, Nathan Reingold, ed. (Washington, D.C.: Smithsonian Institution Press, 1979), 3:385.

154 laudatory sketch of Augustin Fresnel: Ibid., 3:470.

155 "junior officer": Stephen Pleasonton to Acting Treasury Secretary William Hodge, April 25, 1851, RG26, E35, Correspondence Box 7 (1850–1851), #132, NAB.

155 who had been calling for Fresnel lenses: Joseph Gilbert Totten in 1840 proposed using two French lamps (at a cost of $10,000) in the Sandy Hook Lighthouse. *Report of Special Board of Engr Officers to Revising Plan for LH on Flynn's Knoll*, RG26, E16, Misc Records, Box 2, NAB.

155 Abert declined: John James Abert to C. M. Conrad, Secretary of War, April 7, 1851, RG26, E24, Misc Records, Box 2, Communications from officers of the Army Engineering Corps, NAB.

155 "hearty": Susan Fenimore Cooper, "Rear-Admiral William

Branford Shubrick: A Sketch," *Harper's Magazine*, August 1876, 402.

156 "by a hundred cubits": Taken from Thomas Corwin, *Report of Secretary of Treasury on the Coast Survey* (Washington, D.C.: Treasury Department, 1851), 6, quoted in Hugh Richard Slotten, *Patronage, Practice, and the Culture of American Science: Alexander Dallas Bache and the U.S. Coast Survey* (New York: Cambridge University Press, 1994), 90. See also Merle M. Odgers, *Alexander Dallas Bache: Scientist and Educator, 1806–1867* (Philadelphia: University of Pennsylvania Press, 1947).

156 known to keep a bust of Arago: Nathan Reingold, "Alexander Dallas Bache: Science and Technology in the American Idiom," *Technology and Culture* 11, no. 2 (April 1970), 163–177.

156 "You have seen by the papers": William Bradford Shubrick to James Fenimore Cooper, May 2, 1851, in James Fenimore-Cooper, ed., *Correspondence of James Fenimore-Cooper* (New Haven: Yale University Press, 1922), 716.

157 "Having heretofore shown that the cost": Stephen Pleasonton to Thomas Corwin, June 7, 1851, C—No. 4, *Report of the Officers Constituting the Light-House Board* [hereafter *Report*] (Washington, D.C.: A. Boyd Hamilton, 1852), 274.

158 "greatly neglected": Ibid., 21.

159 "an unanswerable argument": Ibid., 24.

159 "beautiful specimen of mechanism": Ibid., 27.

160 government would save $112,185.27 a year: Ibid., 89.

161 "which appears to me": Léonor Fresnel to Léonce Reynaud, July 28, 1851, "Phares Etrangers——Amerique," AN f/14/20908.

161 "with great pleasure, as our light-houses": David D. Porter, July 1851, B—No. 3, *Report*, 207.

161 "had better be dispensed with": H. J. Hartstene, July 18, 1851, B—No. 1, ibid., 212.

161 "I do hope you will": Captain J. C. Delano, November 25, 1851, B—No. 23, ibid., 238.

162 "is the business of an engineer": Ibid., 26.

162 "order out of anarchy and confusion": Ibid., 62.

163 "Could I but put them on a topsail yard": G. W. Blunt, "Lights, etc. (letter to the Editors of the *Courier* and *Enquirer*)," R26, E17B, NAB.

163 "In 1852, the bill for creating": George Alfred Townsend, *The New World Compared with the Old* (Hartford: S. M. Betts, 1869), 387.

164 "a uniform and systematic plan": Thornton A. Jenkins to Henry de Courcy (agent to Sautter), December 6, 1854, RG26 E20, Letters from Light-House Board, vol. 2 (1854–1857), 71, NAB.

164 "notoriously bad": Kevin P. Duffus, *The Lost Light: The Mystery of the Missing Cape Hatteras Fresnel Lens* (Raleigh: Looking Glass, 2003), 20.

165 "the most important on our coast": *Report*, 208 (italics in original).

165 Jenkins had written: Ibid., 383.

165 "special agent": Anna Ella Carroll, *The Star of the West, or National Men and National Measures* (New York: Miller, Orton, 1857), 307.

166 In November of 1852, he asked the Lighthouse Board: Thornton A. Jenkins to I.W.P. Lewis, November 26, 1852, RG26, E20, vol. 1, 43, NAB.

166 recouping six thousand: Carroll, *Star of the West*, 313.

168 "Thus, this marvellous contrivance": George Meade, quoted in Samuel Rhoads and Enoch Lewis, "The Fresnel Light," *Friends' Review* 7 (1854) 87.

168 soon found guilty: Testimony of J. S. Misroon, *United States Congress, Reports of Committees* (Washington, D.C.: A. O. P. Nicholson, Senate Printer, 1856), 21.

168 "The Board hopes": "best results upon the public mind": Thornton A. Jenkins to Henry-Lepaute, September 9, 1851, RG26, E20, Letters sent by Light-House Board, vol. 1 (1852–1854), NAB. Thorton A. Jenkins to Christopher Faye, March 10,

1853, RG26, E20, Letters sent by Light-House Board, vol. 1 (1852–1854), 91, NAB.

169 more money: The prices, which included lamps, were quoted in *Report*, 102.

169 "If the present orders are filled": Thornton A. Jenkins to Lepaute, April 3, 1854, RG26, E20, Letters sent by Light-House Board, vol. 1 (1852–1854), 411, NAB.

169 "Soon, we shall be reduced": Thornton A. Jenkins to Henry de Courcy (agent for Sautter), August 7, 1855, RG26, E20, Letters sent by Light-House Board, vol. 2 (1854–1857), 253, NAB.

170 "any and all": Thornton A. Jenkins to Lepaute, August 6, 1855, RG26, E20, Letters sent by Light-House Board, vol. 2 (1854–1857), 252, NAB.

170 "If Messrs Sautter & Co": Thornton A. Jenkins to Henry de Courcy, August 7, 1855, RG26, E20, Letters sent by Light-House Board, vol. 2 (1854–1857), 253, NAB.

170 In February of 1855, Lepaute asked: RG26, E1, *Journal of the Lighthouse Board*, vol. 3, 145, NAB.

170 "importance to commerce": *Report*, 126.

171 "The large amount of trade": *Report*, 136.

171 in 1850 Congress apportioned generous funds: "Chap. 77—An Act Making Appropriations for Light-Houses, Light Boats, Buoys, &c.," in *Acts and Resolutions Passed at the First Session of the Thirty-First Congress of the United States* (Washington, D.C.: Gideon, 1850), 120.

172 under somewhat suspicious circumstances: Holland, *America's Lighthouses*, 155.

172 "a fruitful source of difficulty, delay, and expense": William Bradford Shubrick, "Report of the Light-House Board, October 31, 1855," Document 54, in *Annual Report of the Secretary of the Treasury on the State of the Finances* (Washington, D.C.: Cornelius Wendell, 1856), 269.

173 "sudden and protracted illness": Hartman Bache to Thornton Jenkins, October 19, 1854, included in "Report of the Light-House

Board," no. 37, *Report of the Secretary of the Treasury on the State of the Finances for the Year Ending June 30, 1854* (Washington, D.C.: Beverly Tucker, 1854), 316.

176 "a tolerably fair job": Hartman Bache to Edmund L. F. Hardcastle, July 11, 1855, Appendix 22, in *Annual Report of the Secretary of the Treasury*, 402.

176 "never less than 45 degrees": Ibid., 403.

176 "only second to an impossible one": Ibid., 403.

177 "The cost in all cases must be large": Ibid., 404.

178 "a wilderness of sand": Robert Louis Stevenson, *The Old Pacific Capital* (New York: Charles Scribner's Sons, 1901), 83.

180 "of a power not less than": Washington A. Bartlett to A. D. Bache, November 29, 1850, Document 21, in *Notices of the Western Coast of the United States, United States Coast Survey* (Washington, D.C.: Gideon, 1851), 53.

180 United States had twice as many lights: "Our Light-house Establishment," *Putnam Magazine* 7 (June 1856), 307.

180 "The prodigious development": Léonor Fresnel to Thornton A. Jenkins, secretary of the Light-House Board, May 7, 1861, in *Papers on the Comparative Merits of the Catoptric and Dioptric of Catadioptric Systems of Light-House Illumination* (Washington, D.C.: Government Printing Office, 1861), 272.

CHAPTER SIX: EVERYTHING RECKLESSLY BROKEN

183 "I am still at my post": Raphael Semmes, *Service Afloat, or the Remarkable Career of the Confederate Cruisers Sumter and Alabama during the War between the States* (London: Sampson Low, Marston, Searle, and Rivington, 1887), 75; for more on Semmes, see John M. Taylor; *Confederate Raider: Raphael Semmes of the Alabama* (Washington, D.C.: Brassey's, 1994).

184 Charleston Harbor: David Detzer, *Allegiance: Fort Sumter, Charleston, and the Beginning of the Civil War* (New York: Harcourt, 2001).

184 "that the coast of South Carolina": Raphael Semmes to secretary of the Treasury, December 20, 1860, cited in George Rockwell Putnam, *Lighthouses and Lightships of the United States* (New York: Houghton Mifflin, 1917), 100.

187 "repair to this place": C. M. Conrad to Raphael Semmes, February 14, 1861, reproduced in Semmes, *Service Afloat*, 75.

188 Edward Tilton: Semmes had been speaking to Tilton only hours before he shot himself in the head. "I have never had anything to shock me so much in the whole course of my experience," he wrote to a mutual friend. Raphael Semmes to John P. Gillis, February 11, 1861, 1-II-102, Papers of John P. Gillis, Delaware Historical Society, Wilmington.

188 "I had barely time": Ibid., 88.

188 "no longer to be thought of": Ibid.

189 "arms, munitions, iron plating": From David George Surdam, *Northern Naval Superiority and the Economics of the American Civil War* (Columbia: University of South Carolina Press, 2001), 1; for more on the blockade, see Robert M. Browning Jr., *Success Is All That Was Expected: The South Atlantic Blockading Squadron during the Civil War* (Washington, D.C.: Brassey's, 2002); Stephen R. Wise, *Lifeline of the Confederacy: Blockade Running during the Civil War* (Columbia: University of South Carolina Press, 1988); W. M. Fowler, *Under Two Flags: The American Navy in the Civil War* (Annapolis, Md.: Naval Institute Press, 2001); Virgil Carrington Jones, *The Civil War at Sea* (Austin, Texas: Holt, Rinehart and Winston, 1962), 3.

190 "The report reached us": *Charleston Mercury*, December 20, 1861, quoted in Douglas W. Bostick, *The Morris Island Lighthouse* (Charleston, S.C.: History Press, 2008), 33.

190 "a very important point": Thornton A. Jenkins to S. P. Chase, October 2, 1861, RG26, E35, Light-House Letters, Box 8, vol. 1 (1860–1861), 188, NAB.

191 "a place which is of such vast importance": Quoted in Kevin P. Duffus, *The Lost Light: The Mystery of the Missing Cape Hatteras Fresnel Lens* (Raleigh: Looking Glass, 2003), 48.

192 "Hatteras Light House is a traitor": W. S. G. Andrews to Governor Henry Clark, Cape Hatteras, July 23, 1861, Civil War: Official Records of the Union and Confederate Navies, NARA, ser. 2, vol. 2, Navy Department Correspondence (1861–1865), 80.

192 "I was desirous of lighting Hatteras light": Commander S. C. Rowan, Report regarding affairs at Hatteras Inlet, N.C., September 3, 1861, U.S.S. *Pawnee*, Civil War: Official Records of the Union and Confederate Navies, NARA, ser. 1, vol. 6, Atlantic Blockading Squadron, 160.

192 "without their knowledge or consent": Commander S. C. Rowan, Report regarding the expedition to Washington, N.C., March 21, 1862, Civil War: Official Records of the Union and Confederate Navies, NARA, ser. 1, vol. 6, Atlantic Blockading Squadron, 150.

192 "I promise protection": Ibid.

192 "take all the property I can find": Ibid.

192 "would go there and frighten": Commander S. C. Rowan, U.S. Navy, to Flag Officer L. M. Goldsborough, U.S. Navy, regarding condition of affairs in the sounds of North Carolina, New Berne, N.C., March 29, 1862, Civil War: Official Records of the Union and Confederate Navies, NARA, ser. 1, vol. 6, Atlantic Blockading Squadron, 178.

193 "unfit to attend": George H. Brown to Confederate Lighthouse Bureau, April 1862, Confederate Treasury Records, NARA, reproduced in Cheryl Roberts, "Letters Reveal Last Journey of the Hatteras Lens but End in a Mystery," *Lighthouse Digest* (March 2000), 8–11.

193 "a good store house": D. T. Tayloe to Thos. E. Martin, April 20, 1862, Confederate Treasury Records, NARA, reproduced in Roberts, "Letters Reveal," 11.

194 Chase to press the War Department for better protection: See War Department's response: Acting Secretary of War Thomas A. Scott to Salmon P. Chase, October 16, 1861, RG26, E35, Light-House Letters, Box 8, vol. 1 (1860–1861), 202, NAB.

194 "may now be lighted": Flag Officer L. M. Goldsborough to Gideon Welles, February 20, 1862, M275, Civil War: Official Records of the Union and Confederate Navies, NARA, ser. 1, vol. 6, Atlantic Blockading Squadron, 636.

196 "a most melancholy example": Flag Officer Samuel Du Pont to Gideon Welles, November 17, 1861, M275, Civil War: Official Records of the Union and Confederate Navies, NARA, ser. 1, vol. 12, South Atlantic Blockading Squadron, 350.

196 "belong to the United States": Flag Officer Samuel Du Pont to Commander J. P. Gillis, November 18, 1861, M275, Civil War: Official Records of the Union and Confederate Navies, NARA, ser. 1, vol. 12, South Atlantic Blockading Squadron, 351.

197 "the apparatus itself ruthlessly destroyed": Flag Officer Samuel Du Pont, Report regarding the condition of light-houses on the Southern coast, April 1, 1862, M275, Civil War: Official Records of the Union and Confederate Navies, NARA, ser. 1, vol. 6, Atlantic Blockading Squadron, 692.

199 "everything recklessly broken": Ibid.

199 "a gang of pirates from St. Augustine": T. Augustus Craven to Gideon Welles, M275, Civil War: Official Records of the Union and Confederate Navies, NARA, ser. 1, vol. 6, Atlantic Blockading Squadron, 207.

200 two hundred strong: C. W. Pickering to Thornton Jenkins, August 28, 1861, RG26, E35, Light-House Letters, Box 8, vol. 1 (1860–1861), 185, NAB.

200 "as citizens of the Confederate States": James Paine to Christopher Memminger, Confederate treasury secretary, October 10, 1861, cited in James D. Snyder, *A Light in the Wilderness* (Jupiter, Fla.: Pharos Books, 2006), 118.

200 "band of lawless persons": T. A. Jenkins to S. P. Chase, September 7, 1861, RG26 E35, Light-House Letters Box 8, vol. 1 (1860–1861), 184, NAB.

200 "The Light being within": James Paine to Madison Starke Perry, Florida governor, cited in Snyder, *Light in the Wilderness,* 118.

201 "shining as the safeguards and symbols": Louis M. Goldsborough to Samuel Du Pont, February 11, 1862, M275, Civil War: Official Records of the Union and Confederate Navies, NARA, ser. 1, vol. 12, South Atlantic Blockading Squadron, 473.

201 two civil officials who did not immediately resign: *New York Times,* March 4, 1862.

203 "He had a power of demonstration": Jefferson Davis, "Autobiographical Sketch," in James McIntosh, ed., *The Papers of Jefferson Davis* (Baton Rouge: Louisiana State University Press, 1974), 1:1xxx.

203 "Home Guards": Mary Renolds to Governor John J. Pettus, November 26, 1861, Mississippi Department of Archives, ser. E, vol. 54, Pettus, reproduced in James Stevens, "Biloxi's Lady Lighthouse Keeper," *Journal of Mississippi History* 36 (1974), 39–41.

205 "in favor of Southern rights as anyone": Manuel Moreno to F. H. Hatch, collector of the Port of New Orleans, February 6, 1861, Confederate States of America, vol. 88, 89, 90, Collectors of Customs Letters, Library of Congress, microfilm reel 45, quoted in Carol Wells, "Extinguishing the Lights, 1861," *Louisiana History* 19, no. 3 (Summer 1978), 300.

207 "I am in this deserted place": Manuel Moreno to F. H. Hatch, March 31, 1861, quoted in Wells, "Extinguishing," 302.

207 "for amusement": Manuel Moreno to F. H. Hatch, June 22, 1861, quoted in Wells, "Extinguishing," 305.

207 "had been strangely overlooked": Raphael Semmes, *The Cruise of the Alabama and the Sumter* (London: Saunders, Otley, 1864), 7.

208 "All my resistance was in vain": Manuel Moreno to F. H. Hatch, July 5, 1861, quoted in Wells, "Extinguishing," 306.

209 Danville Leadbetter: Jeffrey N. Lash, "A Yankee in Gray: Danville Leadbetter and the Defense of Mobile Bay, 1861–1863," *Civil War History* 37, no. 3 (September 1991), 197–218; see also Arthur W. Bergeron, *Confederate Mobile* (Jackson: University of Mississippi Press, 1991).

209 "tumble the Light House": Quoted in Bruce Roberts and Ray

Jones, *Gulf Coast Lighthouses* (Guilford, Conn.: Globe Pequot Press, 1998), 61.

209 Glenn proudly reported: John W. Glenn to Brig. Gen. Danville Leadbetter, Engineer Bureau, Ft. Gaines, Fla., February 24, 1863, Binder 20, Dabney H. Maury Collection, Museum of the City of Mobile, Mobile, Ala.

211 "not deemed advisable": W. B. Shubrick to S. P. Chase, October 31, 1863, Section P, in *Report of the Secretary of the Treasury on the State of the Finances* (Washington, D.C.: Government Printing Office, 1863), 158.

211 "guerilla force from the mainland": Ibid., 156.

212 "The glass concentric reflectors of Fresnel": *Philadelphia Inquirer*, April 26, 1865, quoted in Duffus, *Lost Light*, 140.

212 "I learn that some broken prisms": Montgomery Meigs to W. B. Shubrick, April 26, 1865, M745, Letters sent by the QM Gen, roll 50 (April 3–August 5), 242, quoted in Duffus, *Lost Light*, 142.

213 "the commerce of the world": Quoted in Duffus, *Lost Light*, 153.

CHAPTER SEVEN: THE GOLDEN AGE

215 Chance Brothers: Toby Chance and Peter Williams, *Lighthouses: The Race to Illuminate the World* (London: New Holland, 2008); Harry J. Powell, *Glass-Making in England* (Cambridge: Cambridge University Press, 1923).

215 "If your Birmingham friends": H. Fuller to Fred Burdus, October 3, 1866, BS6/12/1/6/2, Sandwell Community History and Archives, Smethwick, England.

216 "more visitors of all nationalities in a week": Frederick Arrow to James Chance, quoted in James Frederick Chance, *The Lighthouse Work of Sir James Chance, Baronet* (London: Smith, Elder, 1902), 58.

216 "a manufacture from which emanate": *The Illustrated Times*, 1862, quoted in Chance and Williams, *Lighthouses*, 142.

220 "the fitness and sufficiency": Quoted in Michael Schiffer, "The Electric Lighthouse in the Nineteenth Century," *Technology and Culture* 46 (April 2005), 284.

224 "some idea may be had": Quoted in Love Dean, *The Lighthouses of Hawai'i* (Honolulu: University of Hawaii Press, 1991), 42.

224 "This project": *Oeuvres,* 3:420.

226 open mercury bath: C. van Netten and K. E. Teschke first raised the issue of toxic effects, in "Assessment of Mercury Presence and Exposure in a Lighthouse with a Mercury Drive System," *Environmental Research* 45, no. 1 (February 1988), 48–57.

227 Japanese sent a delegation: *The Iwakura Embassy, 1871–1873,* Kume Kunitake, comp. (Princeton: Princeton University Press, 2002), 2:372.

228 originally was not interested: George A. Macbeth, "Light House Lenses," *Proceedings of the Engineers' Society of Western Pennsylvania* 30 (1914), 231.

228 "making Pittsburg light the way": Quoted in "American Made Lenses Best," *National Glass Budget* 29 (March 21, 1914), 4.

229 Second World War: Malcolm Francis Willoughby, *The U.S. Coast Guard in World War II* (Annapolis, Md.: United States Naval Institute, 1957).

229 "the elements of occupation": André de Rouville, Directeur du Service des phares, to chief engineers, July 24, 1940, AN f/14/20848.

230 "It was purely due to Schadenfreude": *Extrait du rapport du subdivisionnaire M. Vernos sur la destruction du Phare du Cap Ferret,* September 21, 1944, AN f/14/20848.

230 "You are not going to blow up the light": Dupouy, *Compte-rendu de l'Etat de balisage,* Morbihan, AN f/14/20848.

231 165 optics had to be replaced: Jean-Marc Fichou, "Les phares français pendant la Seconde Guerre mondiale," *Guerres Mondiales et conflits contemporains,* issue 4, n. 204 (2001), 109–123.

ACKNOWLEDGMENTS

My colleagues at the University of Mississippi have been a resource on everything from Civil War blockades to the Meiji Restoration. Thank you Sara First, Sue Grayzel, Alix Hui, Ari Joskowicz, Jeff Kosiorek, Marc Lerner, John Neff, Peter Reed, Mindy Rice, Sheila Skemp, Jason Solinger, Daniel Stout, Joe Ward, Jeff Watt, Noell Wilson, and everyone else who humored my pharocentric view of history. I also thank the organizers and commentators of sessions where I presented: Elise Lipkowitz, Michael Gordin, Alex Csiszar, Denise Phillips, Andre Wakefield, and Lissa Roberts. I have appreciated the comments and encouragement of Tim Blackwood, Wayne Wheeler, and particularly Thomas Tag, who helped with several illustrations. The archivists and librarians of the Archives Nationales, Archives de l'Académie des Sciences, National Archives, Sandwell Community History and Archives, and Widener Library were all immensely helpful, and a special thanks to Margaret Levitt for her help in the archives. I benefitted immensely from the advice of Nancy Green at Norton, and also thank Ben Yarling for his attention seeing everything through. A final thanks to Bjoern, Juneau, and Clara, for their patience and support. The book is dedicated to my father and favorite engineer, Richard Levitt, who, too, built marvels in the wilderness.

INDEX

Jacket front: (left) The first-order Fresnel lens of the Point of Ayre lighthouse, on Scotland's Isle of Man, was built in 1890 by Barbier & Bernard of Paris, and designed by David Alan Stevenson, the fourth generation of lighthouse engineers in his family. As his cousin Robert Louis Stevenson put it, "there is scarce a deep-sea light from the Isle of Man north about to Berwick, but one of my blood designed it"; (right) the lighthouse at Pigeon Point, California. Its lens, with its distinctive 24 beams of light, was built in 1863 to duplicate the missing lens of North Carolina's Cape Hatteras, hidden by Confederates during the Civil War. Stored when the original lens was found, the duplicate was shipped to California in 1871.